NEAR-DEATH EXPERIENCES

NEAR-DEATH EXPERIENCES

Understanding Visions of the Afterlife

John Martin Fischer

AND

Benjamin Mitchell-Yellin

OXFORD
UNIVERSITY PRESS

OXFORD
UNIVERSITY PRESS

Oxford University Press is a department of the University of Oxford. It furthers
the University's objective of excellence in research, scholarship, and education
by publishing worldwide. Oxford is a registered trade mark of Oxford University
Press in the UK and certain other countries.

Published in the United States of America by Oxford University Press
198 Madison Avenue, New York, NY 10016, United States of America.

Library of Congress Cataloging-in-Publication Data
Names: Fischer, John Martin, 1952– author. | Mitchell-Yellin, Benjamin, author.
Title: Near-death experiences: understanding visions of the afterlife /
John Martin Fischer and Benjamin Mitchell-Yellin. Description: New York, NY:
Oxford University Press, [2016] | Includes bibliographical references.
Identifiers: LCCN 2015038168 | ISBN 978-0-19-046660-2 (hardcover: alk. paper)
Subjects: LCSH: Near-death experiences.
Classification: LCC BF1045.N4 F58 2016 | DDC 133.901/3—dc23
LC record available at http://lccn.loc.gov/2015038168

1 3 5 7 9 8 6 4 2
Printed by Sheridan, USA

CONTENTS

CONTENTS

PREFACE

There is tremendous interest in near-death experiences: their nature, their meaning and their implications for the afterlife. This is understandable. Few topics are of greater interest to human beings than what happens to us after we die, and near-death experiences seem to be a window into the truth on this matter.

These issues have recently been explored in a number of popular books. Three of them, Eben Alexander's *Proof of Heaven: A Neurosurgeon's Journey into the Afterlife* and *The Map of Heaven: How Science, Religion, and Ordinary People Are Proving the Afterlife,* as well as Todd Burpo's *Heaven Is for Real* (also the subject of an extremely successful motion picture), offer an interpretation of near-death experiences as pointing to the existence of an afterlife (as understood in the Christian tradition). Two other books that have reached a wide audience, Jeffrey Long's *Evidence of the Afterlife: The Science of Near-Death Experiences* and Pim van Lommel's *Consciousness Beyond Life: The Science of Near-Death Experience,* call into question the framework of the physical sciences. All five of these discussions of near-death experiences take them as evidence of something supernatural.

We have deep respect for individuals who have had near-death experiences. Indeed, we hope that our serious attention to these remarkable experiences displays our respect. We also respect the deep religious convictions of many who have thought about the significance of these experiences. In these pages we wish to address the *implications* of near-death experiences for questions about the fundamental nature of reality and the relationship between our minds and our brains. Do near-death experiences establish that there is an afterlife? Do they show that our minds can function separately from our brains? Our aim is to explore these questions with the seriousness and rigor they deserve.

Many people have provided us with helpful feedback on the material in this book. In particular, we would like to thank the following people, in alphabetical order: Yuval Avnur, Christie Coy, Richard Coy, Ari Fischer, Tina Fischer, Heinrik Hellwig, John Perry, Patrick Ryan, Eric Schwitzgebel, Allison Solso, and Howard Wettstein. We owe special debts of gratitude to Nathan Ballantyne, Shelly Kagan, and Peter Ohlin for their detailed and extremely helpful comments. The writing of this book was supported in part by a generous grant from the John Templeton Foundation, but the book does not necessarily reflect its views or stances on any of the issues discussed.

NEAR-DEATH EXPERIENCES

Chapter 1

Introduction

In his bestselling book, *Proof of Heaven: A Neurosurgeon's Journey into the Afterlife*, Eben Alexander describes his experience during a seven-day coma. He was presented with "a beautiful, incredible dream-world":

> Except it wasn't a dream. Though I didn't know where I was or even *what* I was, I was absolutely sure of one thing: this place I'd suddenly found myself in was completely real. The word *real* expresses something abstract, and it's frustratingly ineffective at conveying what I'm trying to describe. Imagine being a kid and going to a movie on a summer day. Maybe the movie was good, and you were entertained as you sat through it. But then the show ended, and you filed out of the theater and back into the deep, vibrant, welcoming warmth of the summer afternoon. And as the air and the sunlight hit you, you wondered why on earth you'd wasted this gorgeous day sitting in a dark theater.
>
> Multiply that feeling a thousand times, and you still won't be anywhere close to what it felt like where I was.
>
> I don't know how long, exactly, I flew along.... But at some point, I realized that I wasn't alone up there.... Someone was

next to me: a beautiful girl with high cheek-bones and deep blue eyes. She was wearing the same kind of peasant-like clothes that the people in the village down below wore. Golden-brown tresses framed her lovely face. We were riding along together on an intricately patterned surface, alive with indescribable and vivid colors—the wing of a butterfly. In fact, millions of butterflies were all around us—vast fluttering waves of them, dipping down into the greenery and coming back up around us again. . . .

The girl's outfit was simple, but its colors—powder blue, indigo, and pastel orange-peach—had the same overwhelming, super-vivid aliveness that everything else in the surrounding had. She looked at me with a look that, if you saw it for a few moments, would make your whole life up to that point worth living. . . .

Without using any words, she spoke to me. . . . The message had three parts . . . :

"You are loved and cherished, dearly, forever."
"You have nothing to fear."
"There is nothing you can do wrong."[1]

Alexander's experience transformed his life. And it has resonated deeply with a wide audience (over 2 million copies sold!). This and other near-death experiences (NDEs) understandably hold people's attention. They are deeply meaningful and profound. We want to understand them—to explain what is going on and to comprehend their significance.

1. Alexander (2012a: 40–41).

Our lives are replete with experiences of all sorts—the sight of a tree shedding its leaves in the yard; the sound of a familiar voice on the phone; the smell of coffee in the kitchen. The typical experience occurs during the course of your normal, waking hours. It is the stuff of everyday life. But near-death experiences aren't typical experiences. They are conscious experiences had by people who find themselves in life-threatening situations, or situations they perceive to be life threatening. These are people who were in cardiac arrest or under anesthesia or in comas. But unlike most people in these circumstances, those who have near-death experiences later report having had rich experiences during that time. They report seeing their own bodies receiving CPR or hearing the doctors talking about their operation or visiting with deceased relatives they never knew they had. The hallmarks of a near-death experience include certain kinds of content that fit some particular patterns. Pim van Lommel, a leading researcher on near-death experiences, defines them as involving "a range of impressions during a special state of consciousness, including a number of special elements such as an out-of-body experience, pleasant feelings, seeing a tunnel, a light, deceased relatives or a life review, or a conscious return into the body."[2] This is not the stuff of everyday perception.

Near-death experiences offer a glimpse into the nature of the End to people who come back from the brink. Those who have had them can shed some light on this dark subject for the rest of us. They can speak to something that remains opaque to the rest of us until it's too late. They can make sense of passing away for those

---

2. Van Lommel (2013: 8). For more detailed discussions of the typical patterns of elements of near-death experiences, see van Lommel (2010, 2013) and Holden, Greyson, and Kelly (2009).

of us who remain among the living because they have treaded on both banks of being—life and death—and lived to tell the tale. Near-death experiences are not only useful tools to aid in the human quest to understand death. They are typically also deeply meaningful and transformative for the people who have them.[3] They can help to reorient one's priorities and to come to grips with what matters in life. Near-death experiences offer a glimpse not only into the nature of death but also into the meaning of life.

It is no wonder that the literature on near-death experiences has so thoroughly captured the public imagination. Visiting heaven and reuniting with lost loved ones, revisiting one's life as if it were a movie, witnessing the resuscitation of one's body at the hands of medical professionals: these are incredible, profound things to experience. Naturally, we want to know why they occur and what they mean. Whether due to their transformative effects on those who have them, or simply because their contents are so fascinating, reports of near-death experiences are often quite gripping. These stories need to be told, and the experiences lend themselves to the telling with gusto. Those who come back from the brink and are able to share their experiences are often able to do so in compelling ways, and recent depictions of near-death experiences in books and films have captured the public's imagination.

One of the first great philosophical books, Plato's *Republic*, concludes with the recounting of a near-death experience. Immediately after giving an argument for the immortality of the soul, Socrates relates the myth of Er, a soldier who died in battle but came back to tell what he saw in the other world. Like many other myths in Plato's works, this is meant to supplement Socrates'

3. See van Lommel (2010, 2013) and Holden, Greyson, and Kelly (2009).

philosophical argument and to help instill noble beliefs. In this case, Socrates aims to make one more convincing case for living a just life. This magnificent treatise, filled with arguments for the importance of justice in the good life, rests its case with the tale of one who came back with a message from the other side. Plato's intention is clear. If we remain unconvinced by philosophical argument, the power of myth may yet get the lesson across.

The transformative power of near-death experiences is not lost on us in the 21st century. But Plato's use of Er's story stands apart from the way near-death experiences are presented nowadays. Our contemporaries recount and discuss near-death experiences in books with titles like *Proof of Heaven: A Neurosurgeon's Journey into the Afterlife, Heaven Is for Real: A Little Boy's Astounding Story of His Trip to Heaven and Back,* and *Evidence of the Afterlife: The Science of Near-Death Experiences.* The idea no longer seems to be that near-death experiences are stories that serve to prepare us for rational argumentation. Rather, the idea seems to be that near-death experiences *are* arguments, or perhaps the idea is that there is no need for arguments (traditionally construed). Simply learning of an amazing experience should convince one of the truth about reality. An individual's sincere report of what appear to be supernatural phenomena implies that his or her report is true and that these supernatural phenomena really exist.

Sometimes seeing really is believing. And hearing about what someone saw can be a way of getting at the truth. But it is one thing if an experience can be shared with others and another thing if that experience is not available to them. If we tell you of our recent journey to a beautiful place where everything is lovely, you may want to go and experience it for yourself. When we tell you that you can't, at least not right now, you may wonder why. You may even begin to wonder if this place is real. Are we telling the truth?

Contemporary authors who relate incredible near-death experiences are keenly aware of the need to earn our trust. Eben Alexander, for example, announces his profession in the title of his book and reminds us of it throughout the text. He is a physician and scientist who studies the brain. Who better to tell of what he had seen while in a seven-day coma? Colton Burpo visits heaven during surgery, and his father tells his story in *Heaven Is for Real*. Who but a four-year-old could be more sincere in his report of what he had seen? The conclusion we are supposed to draw is that the testimony of these people is unimpeachable. We can trust them like we do our own eyes. Their words faithfully reveal the way things really are. They are telling the Truth, the whole Truth, and nothing but.[4]

The trouble, of course, is that no matter how hard one tries, it is difficult to please skeptics. Some people just aren't open to trusting the words of others, especially about supernatural phenomena. Even with the Truth laid out right there in front of them, out of the mouths of babes and doctors, those who are a bit more skeptical will remain unwilling to see the light. They will be unable to accept that what looks like a mere story is really and truly a sound explanation. They will be unable to confront the ashes of reality as they knew it, a reality that has been blown apart by the revelations to be found in these profound experiences.

But there are cases of near-death experiences that seem fit to dispel the doubts of any reasonable skeptic. Some who brush with death tell of things they saw and heard happening here on earth.

---

4. In other words, they are not full of "Malarkey"—the unfortunate name of the young man who recently recanted the story about his trip to heaven in his co-authored book, *The Boy Who Came Back from Heaven*. For an account of this boy who—well—did NOT come back from heaven, see Dean (2015).

Their tales are not of trips to heaven, but of trips to the Emergency Room. And we can know that the things they report seeing and hearing really happened because other people—the nurses resuscitating them, the doctors operating on them, their loved ones in the hospital waiting room—all saw and heard these happenings as well. The details of these near-death experiences can be *confirmed*. And that makes them a fit subject for scientific investigation. Perhaps the tools of science can convince the skeptics.

Perhaps. This book is about how we should go about comprehending near-death experiences. They speak to something dear to us—the nature and significance of life and death—and yet we have a hard time figuring out exactly why people have them and what precisely they mean. Near-death experiences confound our attempts at explanation and frustrate our search for significance. How should we respond?

One response to the difficulty presented by near-death experiences is to give up on the prospects of explaining them in the way we explain most other things around us. Though we usually appeal to elements of the physical world and their law-like relations to explain what happens, it may seem as though we cannot make sense of near-death experiences in the usual way. And so we need to appeal to supernatural phenomena, like heaven and souls. Once we do this, the thought continues, we can also better understand their significance. We can fit them into the stories of our lives and our relationships with others. Near-death experiences are meaningful, on this view, because they connect us to a reality that transcends our usual categories of experience.

We have just been describing a view we will refer to as "supernaturalism." A supernaturalist accepts one or both of two claims. The first is that we have access to a supernatural realm—one separate from our ordinary physical world, such as heaven. The second is that our minds are nonphysical and the means by which we come to have experiences do not depend solely on our brains or any other part of our bodies. This second component of supernaturalism is a denial of the doctrine of "physicalism" about the mind. A physicalist will claim that if the brain stops functioning, there will be no consciousness or mental states, because the mind is not separate from our brains. The second element of supernaturalism denies this. For example, a supernaturalist might contend that we have souls and, further, that it would be possible for consciousness to continue even after our brains stop functioning, because the seat of consciousness is found in our souls. Or a supernaturalist might contend that consciousness is a nonphysical force harnessed by our brains but, ultimately, separable from them. This would make room for conscious experiences at times when our brains are not functioning and in places far removed from the location of our skulls. These two components of supernaturalism go together well. If we are able to have experiences in heaven, then consciousness must be separable from our brains and bodies.

Often, proponents of the view that near-death experiences provide support for supernaturalism put the point very strongly. Some contend that near-death experiences show that supernaturalism is *almost certainly* true.[5] For example, after considering several lines of evidence, Jeffrey Long concludes:

> After considering the strength of the evidence, I am absolutely convinced that an afterlife exists.... The NDERF [Near Death

---

5. For instance, Long (2010) and Alexander (2012a) are making some version of this claim. Van Lommel (2010, 2013) suggests this interpretation but may be claiming that near-death experiences show that nonlocalized physicalism is very probably true (we will return to this view in Chapter 9).

Experience Research Foundation] study is the largest scientific study of near-death experience ever reported, and it provides exceptional new scientific evidence for the reality of NDEs and their consistent message of an afterlife. Any one of these nine lines of evidence individually is significant evidence for the reality of near-death experiences and the afterlife. The combination of these nine lines of evidence is so convincing that I believe it is reasonable to accept the existence of an afterlife. I certainly do.[6]

In the chapters that follow, we will consider several of these lines of evidence.

Our method, as intellectual descendants of Socrates, will be to consider the arguments on all sides and to get to the bottom of things, as best we can, using informed, compassionate reason and logic.

We begin our discussion, in Chapter 2, by considering two well-known near-death experiences and how they might be thought to generate an argument against the physicalist view about the mind. Unlike Eben Alexander's near-death experience, described in this chapter, these are examples of near-death experiences that involve reports about our shared world. They are a good place to start our inquiry into the question of whether we have good reason to abandon our common-sense understanding of this world and our place in it.

6. Long (2010: 199, 200–201).

Two Famous
Near-Death Experiences

Pam Reynolds felt herself float out of the top of her head. She hovered in the operating room for some time, witnessing the work of the medical team prepping her body for brain surgery. Some of what she saw surprised her. They shaved only part of her head and prepped her groin area. Why would they do that? Two members of the medical team discussed the difficulty posed by her small arteries. The bone saw had interchangeable blades and, to her ear, an unpleasant sound. Meanwhile, her eyes were taped shut and the respirator pumped on, keeping her anesthetized body alive as her brain was shut down. The electroencephalograph (EEG) electrodes attached to her head picked up nothing. The earplugs molded into her ears emitted rapid, loud clicks, which her brain eventually ceased to register. Finally, the blood was drained from her brain, and with her body temperature at 60 degrees, the doctor repaired her aneurysm. Then they brought her brain and body back to normal.

It was while all of this was happening to her body, Pam would later recall, that she had an incredibly vivid experience—one that would forever change her life. After witnessing the medical personnel prepping her body, she left the operating room for some-place else, bathed in bright light, and encountered deceased relatives who communicated to her without words. Somewhat reluctantly, she eventually returned to her body to live on.[1]

In time, it became clear that the things Pam Reynolds saw and heard in the operating room—the conversation about her arteries, the buzz of the saw and the shape of its blade—really did happen. And yet, we would not expect someone in Pam's condition at the time to be able to see or hear anything at all. How can we make sense of this? A woman saw and heard things while her brain was being shut down for a radical form of brain surgery. She heard a conversation about her small arteries while loud clicking sounds were pumped into her ears and no auditory sensations registered in the measurements taken of her brain. How is that possible?

Other near-death experiences seem similarly incredible. Consider the case of the man with the missing dentures.[2] This patient watched his body from a location above it while under-going CPR during cardiac arrest. When the hospital staff were unable to locate the man's dentures days later, he saw one of the nurses who had attended his CPR and reminded him, the nurse, that he had taken out his dentures and placed them in a drawer, where they must have subsequently been forgotten. The mystery of his missing dentures was solved because this patient was able to tell the hospital staff where they had been placed while he was comatose. And he was able to tell them this information because he had seen the procedures during his cardiac arrest from a position

---

1. For details about and discussion of Pam Reynolds's case, see Williams (2014) and Holden (2009).
2. For discussion of this case, see van Lommel et al. (2001) and van Lommel (2010, 2013). We are not told this man's name.

outside his body. Like Pam Reynolds, this man's near-death experience provided him with details about the world that were confirmed by others. And also like Pam Reynolds, this experience and these details occurred at a time when we would not have expected him to be able to see or experience anything that was going on around him.

It is very difficult to make sense of cases like these in purely physical terms. The usual explanation of an auditory experience invokes sound waves, eardrums, and the brain. But how are we to account, in these terms, for Pam Reynolds's auditory experience of hearing the conversation about her arteries and the buzzing of the saw? It is not at all obvious that we can. She had speakers molded into her ears emitting rapid, loud clicks, but she did not report hearing these clicks. It is not clear how she could have heard the conversation or the whir of the saw without hearing the clicks as well. And yet she did not report hearing clicks, only the conversation and sound of the saw.

But perhaps we can make sense of this in a rather straightforward way. Maybe Pam Reynolds did hear the clicks, but she simply did not remember hearing them. The difference between what she remembered hearing, and so reported hearing, and what she did not remember hearing might have had to do with attention.[3] She attended to the conversation, but not the clicks, and that explains why she remembered the one but not the other. This response, however, does not seem convincing. Let's grant that attention can play a role in determining what you remember and thus what you later report as having experienced. Still, it is not clear how Pam Reynolds could have heard the conversation in the first place. The

3. For a similar claim about the relevance of attention, albeit in a slightly different context, see van Lommel (2013: 19–20).

speed (11 to 33 clicks per second) and volume (90 to 100 decibels) of the clicks would seem to prevent her from hearing the conversation in the first place.[4] So there is a problem to be solved even before we appeal to attention. Given the competing sound of the clicks, how are we to explain Pam Reynolds's experience of hearing a conversation with enough clarity that she was later able to report it in some detail?

The trouble may seem to lie in the assumption that we are to explain Pam Reynolds's near-death experience with the meager resources of the known physical world. Perhaps we should broaden our horizons and admit the possibility that there is more to reality than just the physical stuff we, together with the scientists, normally take to make up the world around us. Maybe auditory sensation is not just a matter of sound waves, eardrums, and neurons. What if we consider the possibility that Pam Reynolds really did exit her body at the top of her head? What if we consider the possibility that she was, for some time during her operation, a disembodied consciousness, something like a floating soul? At first, this nonphysical consciousness hovered outside her body in the operating room; later, it traveled through some otherworldly dimension along with her deceased family members. Finally, it returned once again to her body.

If we take supernaturalism seriously, then we can entertain the possibility that your consciousness is not necessarily dependent on your body—in particular, on your brain—and thus that you

4. See Holden (2009: 198). At this time, the time of the conversation, the brain registered auditory impressions from the clicks. So it is not right to say that there was no sign of the possibility of auditory experience at this time. The puzzles are these: How is it possible that she even heard the conversation at all? And how is it possible that she did not experience hearing the clicks? For discussion of the possibility that Pam Reynolds could hear in spite of the earphones and clicks, see Woerlee (2011).

can continue to have experiences when separated from your body. It seems that in accepting supernaturalism we afford ourselves the explanatory resources to make sense of near-death experiences like the one reported by Pam Reynolds. She really did hear the conversation and the buzz of the saw, really witnessed what the medical team was doing to her body, because she, as a nonphysical consciousness, had separated from her body for some time. We might explain the case of the man with the missing dentures in a similar way. He really did see the nurse take out his dentures and put them in the drawer while his body was undergoing CPR, because he was no longer located in that body. He was floating near the ceiling, soaking it all in.

Reflecting on these cases suggests that we should take supernaturalism seriously. Indeed, it may even suggest something stronger, that the view that everything is physical is unreasonable. In other words, near-death experiences like those of Pam Reynolds and the man with the missing dentures seem to support supernaturalism and tell against physicalism.

The main issue here turns on time. The key aspect of these near-death experiences is that they involve events that are confirmed by others to have occurred at the time when they seemed to the subjects of these near-death experiences to have occurred. The conversation that Pam Reynolds overheard seemed to occur during her surgery prep, and the medical staff confirmed that it occurred at that time. The nurse remembered taking out the man's dentures during his resuscitation, which was the same time it seemed to the man that he was watching things from outside of his body. These near-death experiences appear to support the claim that at least some near-death experiences are real conscious experiences that occurred at the time they are claimed to have occurred—a time at which it would be difficult to make sense of conscious experience in physical terms. They seem to be real conscious experiences, because they involve real events involving the bodies of their subjects. They seem to have happened at the time they are claimed to have happened, because these events are uniquely related to that particular time. The conversation about Pam Reynolds's veins took place only once, during the time she was under anesthesia; the man's dentures were removed and put into that drawer only once, during the time of his resuscitation.

The apparent timing of these near-death experiences is key to an argument against physicalism because it appears to block some fairly obvious physical explanations. For example, it is not plausible that these people are confusing a past memory for a later experience. If they were ever to have seen and heard these things, as they report, it must have been at these particular times—times during which their brains did not appear to be capable of anything like normal functioning. The fact that these experiences match what happened at a specific time also appears to preclude hallucinations as a plausible explanation. We do not think of hallucinations as representing reality, and these near-death experiences did reflect what really happened. So the claim that Pam Reynolds or the man with the dentures were hallucinating and that this explains why they reported seeing themselves from a position outside their bodies doesn't seem to hold water. This explanation is even less plausible in light of the unique contents of their reported experiences. Pam Reynolds did not expect to see much of what she saw. So there seems to be no good explanation for why she would have had a hallucination including these specific features.

The cases of Pam Reynolds and the man with the missing dentures appear to pose a powerful challenge to the claim that the mind is entirely physical. While the functioning of their bodies, and especially their brains, was severely impaired, these people had experiences that included events that really happened. We know these events happened because the medical personnel working to keep these people alive confirmed that they really happened. And we know that these events happened at a time when we would not expect these patients to be capable of any conscious experiences at all. The normal ways of explaining what one sees and hears, in physical terms, do not seem apt in these cases. Thus, explaining these near-death experiences seems to require appealing to non-physical factors, such as a soul or disembodied consciousness.

On the basis of these cases, we see the outlines of a convincing argument against physicalism and in favor of supernaturalism. First, we have confirmation that at least some near-death experiences are real conscious experiences had at the time they are claimed to have been had. Second, any complete explanation of these near-death experiences must appeal to the nonphysical. It follows from these claims that not everything is physical, and, in particular, that the mind is not entirely physical. This argument is indeed very tempting.[5] But, as we shall argue in Chapters 3 and 4, the temptation should be resisted.

---

5. See Mitchell-Yellin and Fischer (2014) for an extended critical discussion of the version of this argument found in van Lommel (2013).

Chapter 3

When Exactly Do Near-Death Experiences Take Place?

It's about time. We don't doubt the sincerity of many of those who report having near-death experiences, nor do we doubt that many near-death experiences are real experiences. We do, however, think there is room to doubt that the subjects of near-death experiences really had these experiences at the *time* it seemed to them that they had them. For all we know, it is possible that in every case, including the cases of Pam Reynolds and the man with the missing dentures, they each had an experience *as if* it were at one time when, in fact, the experience was really at some later time. This point about time casts doubt on the first premise in the argument against physicalism sketched at the end of Chapter 2: the claim that we have confirmation that at least some near-death experiences are real conscious experiences occurring at the time they are claimed to have happened.

This kind of phenomenon is not uncommon. A dream may seem as if it lasted for hours when in fact the processes in the brain lasted only a short time (perhaps the time just before you woke up). And one can wake up with the impression that one's dream occurred at the beginning of the night, when it actually occurred in the moments before dawn. Hallucinations, even those induced by certain drugs, provide another example of the apparent timing of an experience diverging from the actual time it took place. The duration of a hallucination may seem much longer or shorter than it actually is. An LSD trip may seem to last for only a few minutes when in reality it lasted for hours.

The possibility of a disparity in actual versus perceived timing of these kinds of experiences, in itself, presents no challenge to physicalism. Consider what the physicalist might say about dreams. Let's suppose that brain activity supports and even entirely produces your nightly experiences of dreaming. It's possible that this brain activity takes place in the few minutes before you awaken, as your brain is ramping up for the day, even though it might seem to you as if your dreams took place during an extended period of time. It may seem as if a dream lasted all night, even if it occurred only a short time before you woke up. Just as the apparent timing of a dream and the timing of its causes may be different, so the apparent timing of a near-death experience may be different from the timing of its causes. A near-death experience that one experienced as occurring during one time could have taken place at another.

Moreover, the possibility of a disparity between the actual and perceived timing of an experience has implications for which causes seem plausible. It's possible that the actual timing of a near-death experience was not during the throes of a near-death episode but instead at a time when robust brain activity was clearly possible—for instance, when the person's brain activity was ramping up after a serious medical procedure. The actual timing of this experience would then pose no difficulty for explaining it in terms of brain activity. And this remains the case even if the apparent time of the experience was one during which brain functioning was severely impaired.[1]

One last point. The apparent timing of the events depicted in an experience, such as a dream, may be different from the actual timing of those very events. For example, you can relive a moment from your youth by dreaming about it. This memorable event may have actually happened decades before you experience it (again) in your dream. And yet your dream presents it as happening *now*. It's possible for the events depicted in a dream to seem as if they occurred at one time when in fact they occurred at some other time. There seems to be no good reason to think that the same cannot be said for near-death experiences.

This makes room for explanations of certain near-death experiences that weren't apparent at first glance. Consider the man with the missing dentures. He reported seeing what was happening to his body while undergoing CPR and was able to accurately describe aspects of his treatment and the appearances of those treating him. This makes it seem as if he must have had a conscious visual experience during the time at which he was in cardiac arrest. And this was, indeed, how he experienced things. It seemed to him as if he saw these things at the same time that he was in a life-threatening state. But once we recognize that the time at which it seemed to him as if he saw these things might not have been the time at which he actually had the experience of seeing them, we open our minds to an alternative understanding.

It is possible that the man with the missing dentures constructed his unique visual experience from nonvisual sources while he was undergoing CPR. He was in the hospital for several days, so he presumably became familiar with the faces of many of

1. For a nice statement of this claim, see Sacks (2012b).

the medical staff working there. It's possible that he had some kind of visual experience while in the hospital and then interpreted back into it accurate representations of the faces of some of those who gave him CPR. It is also possible that he was able to report what happened to his dentures without ever witnessing what in fact did happen to them.[2] Perhaps he saw other patients' dentures removed and placed in similar locations. Or perhaps, while unconscious, he registered the feeling of having his dentures removed and the sound of the drawer being opened and the dentures being placed in it. After regaining consciousness, he might have (subconsciously) pieced together these sensations, registered while he was unconscious, into the account he gave to the nurse. It may have been incredibly felicitous that the account he pieced together at a later time, while not quite accidentally true, turned out to be accurate. None of these possibilities is ruled out by the fact that the content of this man's near-death experience was corroborated by the nurse.[3]

We are contesting the claim that near-death experiences show that it is *extremely likely* that supernaturalism is true. We are not attempting to *prove* that the sorts of alternative scenarios described above actually took place. Rather, our point is that, given these possibilities, it does *not* seem extremely likely that the only or best

2. Interestingly, the account van Lommel provides of this man's near-death experience is out of the nurse's mouth. Moreover, the nurse does not attribute to the patient the report of seeing him (the nurse) handle his dentures. The nurse says that the patient reported seeing his body undergoing CPR and quotes the patient as telling him about what he did with his dentures. But he does not quote the patient as saying that he saw the nurse remove his dentures and put them away. See van Lommel et al. (2001: 2041) and van Lommel (2013: 18–19).

3. This is all, of course, to treat the nurse's corroboration as genuine. From the account as given by van Lommel, it is possible that the nurse conformed his memory of events to the report given by the patient days later. This raises issues about the methods by which the contents of reported *near-death experiences* are to be verified.

explanation of these experiences implies supernaturalism. At the very least, we should give the possibility of physical explanations serious attention—more attention than it receives in the literature on near-death experiences.

We think that attention to various neglected possibilities opens up interesting ways of interpreting Pam Reynolds's much discussed near-death experience as well. She reported hearing the conversation about her vein size, but she did not report hearing the clicks that were pumped into her ears at higher decibels than occur in a normal conversation. In these circumstances, it would be incredible if she were able to hear the conversation by normal means. It is even more incredible, given that she did not report hearing the clicks.[4] It does not, however, seem unreasonable to suppose that these sounds all registered somewhere in Pam Reynolds's brain and that only some of them were brought to her conscious awareness after her brain functioning was restored. That is, she may have, at some later time, become consciously aware of auditory impressions that her brain received, at some earlier time, and that she did not, at that earlier time, consciously experience. And it is possible that she did not report the sounds of the clicks because they did not, at the later time, come to her conscious attention, even though the sounds of the conversation did.

We are not proposing something mysterious here. This kind of thing happens all the time. You are driving home from work, just as you do every day, following your usual route, listening to your usual radio station, on autopilot. When you get home, your wife asks you about the horrible accident in front of the post office. It's all over the evening news. At first, you stare at her blankly. What accident? Then the flashing lights and lane closure come back to

---

4. See Holden (2009: 198–199); and also Woerlee (2011).

you. You remember seeing it now, though you weren't conscious of doing so at the time. You were lost in thought or absorbed in the song on the radio. Now that you think of it, the accident must've been bad. Lots of commotion. Of course, you don't recall everything you passed by on the way home, just some of what is relevant, given your wife's question. If she asked about the price of gas at the station down the street, you might be able to recall that too. But she doesn't, and you don't. Even if the numbers on the sign registered, you have no reason to remember them.

Essentially the same thing might have occurred in Pam Reynolds's case. Events registered in her brain under her conscious radar, and she later recalled some of them because they were relevant. But there is a crucial difference between the two cases. Presumably, you have experiences while driving, even in autopilot mode. What you recall seeing in front of the post office was not just something that registered in your brain and later came to conscious awareness. It was an experience that you were not conscious of having at the time you had it. But we need not assume that Pam Reynolds had any experiences while her brain was being prepped for surgery. Rather, we are suggesting that her brain registered certain auditory sensations at that time. And we are distinguishing between, on the one hand, having an auditory experience and, on the other, auditory sensations registering in one's brain. While the case of driving home on autopilot shows that there is something reasonably familiar about our suggested interpretation of Pam Reynolds's case, the two cases are not exactly similar. They both involve remembering having had an experience that did not register in consciousness at the time that the events represented in the experience occurred. But, we are suggesting, the case of Pam Reynolds might not involve her having had an experience at all at the time the events occurred.

We are not raising the possibility that Pam Reynolds was conscious at the time the conversation took place. Rather, we are raising the possibility that, even though she was unconscious, auditory impressions may still have registered, and they could have come to her conscious awareness later.[5] Perhaps, as seems possible in the case of the man with the missing dentures, her report was cobbled together after the fact from data she acquired but was not consciously aware of at the time she acquired it. It seems possible that she may have constructed the conscious near-death experience she later reported out of data that registered in her brain but was not, at the time it registered, something she experienced. And yet this cobbled-together-after-the-fact experience may have represented events that actually happened.

What the case of Pam Reynolds illustrates is the possibility that someone can report a conscious auditory experience at a later time than the auditory system actually received it.[6] This would be an example of a conscious experience being generated in an unusual way that nevertheless involves the normal physical mechanisms. Because the usual mechanisms are implicated in the process, the accuracy of the experience would not seem mysterious. We can make sense of why one might report hearing sounds that originated in actual events that made auditory impressions on one's brain. It may be mysterious why these impressions did not register in consciousness immediately. Why did she become conscious of these sounds only at some later time? Or it may be a mystery why some impressions registered in consciousness while

5. Thus, the point we are making here is different from the one made by, for example, Woerlee (2010). For this reason, it is not subject to the same critique, offered in response to Woerlee's article, by Smit and Rivas (2010).

6. Greyson et al. (2009: 229–230) cite some other authors who make a similar point with respect to out-of-body experiences.

others did not. We might wonder why Pam Reynolds reported the conversation but not the clicks. But these are different mysteries than why the conscious experience matched reality. Though they may be strange, physical explanations like these do not seem impossible. Moreover, and this is a point we will return to in later chapters, they certainly do not seem any stranger than explanations that invoke a disembodied consciousness. Faced with the choice of explaining Pam Reynolds's near-death experience in terms of the unusual operation of familiar physical mechanisms or in terms of a disembodied consciousness, it seems reasonable to prefer the former.

These potential physical explanations of the near-death experiences of Pam Reynolds and the man with the missing dentures bear on the argument against physicalism we sketched in Chapter 2. That argument begins with the claim that we have confirmation that at least some near-death experiences are real, conscious experiences occurring when their subjects thought they had occurred. Cases like those of Pam Reynolds and the man with the missing dentures are supposed to provide support for this claim. We have called this into question by suggesting ways in which we might understand how these experiences came about, without assuming that they occurred when they seemed to. We have suggested, in particular, that it may have been possible for these people's brains to register information at one time, and then for them to later remember having a conscious experience based on this stored information. Perhaps Pam Reynolds's brain registered auditory sensations encoding the conversation about her veins at the time the conversation took place. At this time, her brain was being prepared for surgery and so was not fully functioning, but it was also not yet drained of blood and so not yet fully nonfunctioning either. Perhaps, at a later time, she then pieced together a conscious experience of hearing this conversation, and it seemed to her as if she had this experience at the time the events took place. If this is a live possibility, as we have argued it is, then her report of having had an experience at a time when her brain was not fully functioning does not support the claim that she was capable of conscious experience at that time. The argument against physicalism is off to a bad start.

But we recognize that our proposal may not put the issue to rest. If Pam Reynolds's brain was really not functioning during her surgery, then how could it record auditory sensations at all? Even if we suppose that her brain was not functioning at a level that would support conscious experience, our proposal only makes sense on the assumption that her brain was functioning to some degree. But the description of the case involves the claim that her brain was not functioning at all. We will go on to consider the merits of this response in Chapter 4.

Chapter 4

Must Near-Death Experiences Be Explained by the Supernatural?

The second premise of the argument against physicalism states that any complete explanation of these near-death experiences must appeal to the nonphysical. This is a very strong claim, indeed. It does not say, for instance, that there appears to be good reason to appeal to the nonphysical. It makes the bold statement that we *must* do so. And it does not say simply that we must do so in some, or even many, cases. Rather, this claim brashly asserts that we must do so in *all* cases of near-death experiences. This is indeed the claim of proponents of the argument against physicalism, such as Eben Alexander and Jeffrey Long (among many others). Reflection on the progress of science and our understanding of the physical world should call this claim into question.

The progress of science is evident from even our most cursory familiarity with the history of ideas. We once thought that the world was flat and that the sun orbited around us. We now know otherwise. The list of things we used to take as fact that we now regard as fiction is long—and growing! Science is an enterprise built on a method of testing claims about the observable world around us. It is the best available means of pursuing knowledge of the physical world, in large part because it is so good at revealing our mistaken pronouncements on matters of fact. And there is no reason to think that it will be any kinder to our current body of supposed knowledge than it has been to the bodies of supposed knowledge that structured our ancestors' understanding of the world. Though we have their mistakes to learn from, we have no reason to think that we have gotten to the absolute bottom of things. Understanding the physical world is an ongoing project, and we should expect to continue gaining new insights all the time. Whatever confidence we have in our current state of understanding should be tempered by acknowledgment of our place in the ongoing chain of scientific advancement.

One further general point follows quickly on the heels of this statement about scientific progress. We should be appropriately wary, not only of what we take to be fact, but also of our expectations about what might replace our current understanding. We may not be in a particularly good position to predict where the march of knowledge will lead. Any predictions we make now are necessarily based on the body of purported facts we have at our disposal. Because what we take to be facts may be wrong, the predictions based on them might be wrong as well. Just as we should temper our confidence in our current body of knowledge, we should also temper our confidence in our expectations of what will replace it.

These reflections should cast serious doubt on the claim that any complete explanation of near-death experiences must appeal to the nonphysical. This requirement reflects the same hubris history counsels against. Even if our current body of scientific knowledge is not up to the task of explaining near-death experiences in physical terms, this does not mean that *no* such explanation is possible. We would do well to learn from the past and avoid such strong pronouncements. We should not dismiss, out of hand, the possibility that future progress may allow us to provide scientific explanations of things we cannot currently make sense of within the reigning scientific paradigm.

There are some particular ways in which we can expect that our scientific understanding of consciousness, and hence our ability to explain it in physical terms, will improve over time. Begin with our methods of measurement and observation of brain activity. Perhaps most of us are familiar with the various acronyms that name the methods by which we are currently able to measure the functioning of our brains—such as PET (positron emission tomography, which measures positron emissions), EEG (electro-encephalogram, which measures electrical activity), and fMRI (functional magnetic resonance imaging, which indicates blood flow). We use these tools quite often and for various purposes. But we were not always able to do this. Our techniques have grown in number and improved in quality, especially over the past few decades. Given the many pursuits in which these techniques are valuable, we should expect them to continue to improve. No doubt we will devise more precise versions of current technologies and invent new technologies as well.[1] Just as the methods we currently employ have allowed us to learn a great deal more about the brain than we ever knew or expected to know, the methods we will employ in the future should reveal new insights. In some instances, we may be able to provide educated guesses about what this new knowledge will be, but we should expect surprises as well.

---

1. The catalyst is not just private enterprise but also the public purse. Recently both the United States and the European Union have pledged to devote significant resources to the development of new technologies for understanding how our brains work. For more information on the US BRAIN Initiative, see http://www.nih.gov/science/brain/; for more on the EU Human Brain Project, see https://www.humanbrainproject.eu/.

We can't predict where our efforts will lead as we learn more about the way our brains work.

It is possible, and seems quite likely, for us to learn that our current methods for measuring brain activity are shallow, capturing activity only above a certain threshold. This raises the possibility that we may find our current methods incapable of capturing all brain activity, or even all brain activity relevant to conscious experience. We may find that in some of those patients we thought had lost all brain function, in fact their brains were functioning at a level undetectable by our current methods. In particular, it is possible that this is true of patients who had near-death experiences. This is a good reason for skepticism about the bold claim that a physicalist explanation of near-death experiences is impossible.

Some recent research into brain processes at the time near an individual's death provides further reason for thinking it is hasty to conclude, from the perspective of our current knowledge, that physical explanations of near-death experiences are not forthcoming. There is some evidence that rather than being less active during cardiac arrest, the brain may be more active at that time than it normally is.[2] This suggests that it is incorrect to assume that simply because one is in cardiac arrest or some other near-death state, one's brain is not functioning in a manner that would plausibly support conscious experience. This is one concrete example of how scientific progress can come to undermine common assumptions about the physical processes potentially relevant to explaining near-death experiences. It suggests that we should not assume

2. See Borjigin et al. (2013). This study was focused on the brains of rats, but it suggests that the state of activity of our own brains when we are near death may not be what it is commonly assumed to be. Of course, further research is needed, both to draw the connections between these results and human brains and to specify exactly what to say about brain activity during cardiac arrest and other near-death states.

that the brains of cardiac arrest patients, such as the man with the missing dentures, are not functional.

To summarize, there is a seductive line of thought that goes like this. Some people report having conscious experiences of events that really did occur at a time when their brains were apparently not functioning. Because physical explanations of conscious experience will appeal to brain activity, there are no physical explanations of these experiences. We must appeal to something nonphysical to explain these phenomena. Thus, physicalism is false—the mind is not entirely physical.

This line of thought is faulty for several reasons. First, the assertion that these people had conscious experiences at times when their brains were not functioning (or not functioning well enough to support conscious experience) is not warranted. One worry about this assertion has to do with the time of the experience. It is at least possible that Pam Reynolds and the man with the missing dentures pieced together the conscious experiences constituting their near-death experiences at some time *after* the events represented in their experiences actually occurred. It is simply not clear that these are cases of conscious experiences had at a time when brain functioning was impaired. It is possible that these patients had the reported experiences after they had fully recovered from their medical interventions or as their brains were coming back on line, so to speak. But it *seemed* to them as if they had these experiences at an earlier time, while they were in the thick of their respective medical procedures. The apparent timing of these experiences may seem right to them, and to the rest of us, because the contents of their experiences represent events that occurred at this earlier time. But this appearance may be misleading.

There is a second reason to worry about the assertion that these were conscious experiences occurring at times when these patients' brain functioning was impaired. Even if these people's brains were not functioning at levels capable of supporting conscious experience at certain times, their brains may still have been functioning to some degree. And this makes room for the possibility that their brains were capable of recording information that the people could later recall in the form of remembered conscious experiences. This is bad news for the critic of physicalism. For if these patients' brains were functioning when they had their near-death experiences, then there is no special problem of accounting for these experiences in physical terms. There is a parallel here with the situation in which brain activity that underwrites dreaming takes place just prior to an individual's waking up, although the dream seems to take place over a longer time period. In general, we should be cautious about making claims about when a conscious experience took place on the basis of when it *seemed* to the subject of the experience to take place. There is no reason to throw this caution to the wind in cases of near-death experiences.

The final reason this line of thought is faulty is that it does not adequately take into account the progress of science. It relies on unwarranted confidence in our ability to measure brain activity. The critic of physicalism asserts that at the times during which it seemed to these people that they were having conscious experiences representing what was happening to their bodies in the hospital, there was insufficient brain activity to support conscious experience. Perhaps this seems like a reasonable thing to say, given our current knowledge about brain activity and our current ability to measure it. But though our methods may appear advanced from our present perspective, this does not mean that they won't seem crude when we look back at them after new discoveries are made. It seems quite reasonable to allow for the possibility that future technologies will be able to detect brain functioning in cases where current technologies tell us there is none. So it is at least possible that the brains of near-death experiencers really are active at the times it seems to them that they have conscious experiences. Again, there may be no special problem accounting for their near-death experiences in physical terms. Hence, it is hasty to conclude, as the second premise of the argument against physicalism claims, that we *must* appeal to the nonphysical in order to explain these near-death experiences.

So much for the argument in favor of supernaturalism and against physicalism sketched at the end of Chapter 2. If that argument were the only seductive line of thought that moves from a claim about near-death experiences to the conclusion that supernaturalism is true, then we could end this book here. But there are other reasons to think that near-death experiences support supernaturalism. Indeed, one of the remaining arguments for supernaturalism on the basis of near-death experiences amounts to a less ambitious version of the argument we have been considering. Rather than claiming that there are *no* physical explanations for the phenomena and that we *must* appeal to the nonphysical in order to explain near-death experiences, this less ambitious argument appeals to the thought that a supernatural explanation of near-death experiences is *better* than a physicalist one.

We will tackle this less ambitious argument head-on in Chapters 8 and 9. But before we assess the relative merits of physicalism and supernaturalism, let's look at several other lines of thought that appear to provide very compelling support for the supernaturalist interpretation of near-death experiences. We begin, in Chapter 5, with reasoning based on the exceptional vividness of near-death experiences.

Are Vivid Experiences More Accurate Because They Are Vivid?

Near-death experiences are not your usual conscious experiences. One way in which they are unique is that they typically involve especially lucid thoughts and perceptions.

> Near-death experiencers often describe their mental processes during the NDE as remarkably clear and lucid and their sensory experiences as unusually vivid, surpassing those of their normal waking state. An analysis of 520 cases in our collection showed that 80 percent of experiencers described their thinking during the NDE as "clearer than usual" or "as clear as usual." Furthermore, in our collection, people reported enhanced mental functioning significantly more often when they were actually physiologically close to death than when they were not.[1]

The impression one gets from reports of near-death experiences is that they are not like dreams or even like the normal experiences of waking life. They reveal things in an especially clear manner.

---

1. Greyson et al. (2009: 229).

Often near-death experiences are presented not just as conscious experiences of things in the world but as exceptionally vivid experiences that allow us to access reality in a special way. Just as some things seem too good to be true, near-death experiences seem too vivid to be normal. Because of this, one might think that whatever explains these experiences is not the same as what explains everyday cases of seeing and hearing what is around us. Something else seems to be at work in these cases. So, the thinking goes, even if we can explain everyday conscious experience in physical terms, we cannot explain near-death experiences in the same way because they are so vivid.

This pattern of thought is on display when Long writes: "People who have had near-death experiences often describe enhanced and even supernormal vision. This is powerful evidence that something other than the physical brain is responsible for vision during NDEs."[2] According to Long, the lucidity of near-death experiences indicates that they must be generated by nonphysical mechanisms. It supports one tenet of supernaturalism.

Others have taken the lucidity of near-death experiences to indicate the truth of the other tenet of supernaturalism—that they provide access to a nonphysical reality. A good example of this sort of approach is found in Alexander's account of his own near-death experience, which we referred to in Chapter 1.[3] During his coma, Alexander experienced a wondrous seven-day episode, which he has been able to recount in immense detail. The contents of this near-death experience cannot be independently confirmed, since it did not involve events that could have been witnessed or corroborated by others. He describes what he saw as a "beautiful,

2. Long (2010: 61).
3. See Alexander (2012a, 2012b).

incredible dream world." But then he continues: "Except it wasn't a dream. Though I didn't know where I was or even what I was, I was absolutely sure of one thing: this place I'd suddenly found myself in was completely real."[4]

The proclamation that this experience was "real" is meant to have some gravity. But while we are more than willing to grant that Alexander's experience was real in one sense, there is good reason to doubt that it was real in a different sense. It is one thing to claim that a certain experience actually occurred and quite another thing to contend that it represents things that are in fact true. People dream. People sometimes really do have experiences that do not at all represent the way things are. Even when not dreaming, they sometimes see pink elephants. But this experience cannot truly represent things as they are because there are no such things as pink elephants. We need some way of talking about experiences that allows us to mark the distinction between seeing pink elephants and seeing gray ones. We shall use the term "accurate" throughout the remainder of the book to mean "truthful" or "corresponding to external reality." An *accurate* experience *depicts reality as it really is.*[5] Not all real experiences are accurate. Hallucinations, illusions, dreams, delusions—these may all be real experiences. People really do have them. But they are not accurate experiences. They do not correspond to reality.

---

4. Alexander (2012a: 39).

5. In the following passage, Long fails to make the distinction between the different ways in which an experience can be real:

> Respondents describe these experiences in a variety of ways, calling them "unspeakable," "ineffable," "unforgettable," "beautiful beyond words," and so on. More than 95 percent of the respondents feel their NDE was "definitely real," while virtually all of the remaining respondents feel it was "probably real." Not one respondent has said it was "definitely not real."
>
> (Long 2010: 202–203)

Might Alexander in the above passage simply be asserting the reality of his experience rather than its accuracy? No, it is absolutely clear that Alexander infers (in part) from the clarity and vividness of his experience that it represented things in external reality as they truly are. Consider, for instance, this passage from a chapter entitled "The Ultra-Real":

> But while I was in a coma my brain hadn't been working improperly. *It hadn't been working at all.* The part of my brain that years of medical school had taught me was responsible for creating the world I lived and moved in and for taking the raw data that came in through my senses and fashioning it into a meaningful universe: that part of my brain was down, and out. And yet despite all of this, I had been alive, and aware, *truly aware,* in a universe characterized above all by love, consciousness, and reality. (There was that word again.) There was, for me, simply no arguing this fact. I knew it so completely that I ached.
>
> What I'd experienced was more real than the house I sat in, more real than the logs burning in the fireplace.[6]

There is a common thread in this passage from Alexander and the earlier one we quoted from Long. They both involve a move from a claim about the striking lucidity of a near-death experience to a further claim that it truly depicts external reality. Long contends that the near-death experience is evidence of the nonphysical nature of the mind; Alexander holds that the other-worldly

Of course, one who reports having an experience would be inclined to think it is definitely or probably real in the first sense we distinguish in the text. But even real experiences in this sense may be unreal in the second sense we introduce. People really have hallucinations, and they may be aware that what they are hallucinating is not accurate.

6. Alexander (2012a: 129–130).

content of the near-death experience is true. Both of these moves are problematic. Why the lucidity of an experience should imply a nonphysical mechanism in the mind is mysterious. Similarly, it is puzzling that one should leap from the lucidity or clarity of one's experience to the truth of what is represented in that experience.

Once again, consider dreams. Dreams neither present things the way they actually are, nor do people generally suppose that they are underwritten by nonphysical mechanisms. Yet dreams can be lucid. There is no serious debate about the nature of lucid dreams. No one seriously contends that the lucidity (or vividness) of a dream supports the conclusion that the contents of that dream match external reality. Nor is there serious consideration of the claim that lucid dreaming supports the conclusion that consciousness is produced by a nonphysical mechanism. But if we do not take the lucidity of dreams to support claims about the accurate depiction of reality and the operation of nonphysical mechanisms, then why should we take the lucidity (or vividness) of near-death experiences to support these claims? Reflection on the case of lucid dreams should give authors like Long and Alexander pause. Yet rather than pause, they both plunge. They fall into the trap of taking the contents of certain experiences at face value because of the manner in which they are presented.

Are near-death experiences unlike dreams in that they are *more* vivid? Perhaps. But this doesn't help. If an experience doesn't necessarily correspond to reality just because it is vivid, it is hard to see how the degree of vividness would make a difference. If a vivid experience can be inaccurate, why not think that an *incredibly vivid* experience might also be inaccurate? The trouble is that the manner in which an experience presents its contents is not a reliable guide to the truth of those contents. Doubling down on the quality of the experience simply does nothing to help ensure that it is accurate.

We should not infer the accuracy of an experience from its vividness. We should also be wary of moving from the claim that an experience is especially lucid to the claim that it is evidence of nonphysical mechanisms. The mere fact that they present lucid contents does not persuade one that dreams are produced by nonphysical mechanisms. Why would things be different with respect to near-death experiences?

In his fascinating book, *Hallucinations*, Oliver Sacks describes many instances of vivid, lucid hallucinations and emphasizes that it is a mistake to suppose that all hallucinations have a dreamlike quality. He writes:

> Perceptions are, to some extent, shareable—you and I can agree that there is a tree; but if I say, "I see a tree there," and you see nothing of the sort, you will regard my "tree" as a hallucination, something concocted by my brain or mind, and imperceptible to you or anyone else. To the hallucinatory, though, hallucinations seem very real; they can mimic perception in every respect, starting with the way they are projected into the external world.[7]

He notes a bit later:

> Hallucinations often seem to have the creativity of imagination, dreams, or fantasy—or the vivid detail and externality of perception.[8]

---

7. Sacks (2012a: ix).
8. Sacks (2012a: xiii).

In discussing research on hallucinations experienced by subjects who had undergone various modes of sensory deprivation, Sacks writes:

> Several subjects spoke of the brilliance and colors of their hallucinations; one described "resplendent peacock feathers and buildings." Another saw sunsets almost too bright to bear and luminous landscapes of extraordinary beauty, "much prettier, I think, than anything I have ever seen."[9]

It is clear from Sacks's case studies that hallucinations may be especially vivid and lucid. And yet consider the various indisputably physical factors that can cause them: "sensory deprivation, parkinsonism, migraine, epilepsy, drug intoxication, and hypnagogia [the state just prior to falling asleep]."[10] It is evidently mistaken to suppose that we can use the lucidity or vividness of an experience to infer that it must be caused by nonphysical mechanisms.

However, maybe focusing on hallucinations and dreams is barking up the wrong tree. Near-death experiences, unlike even lucid dreams, are typically "coherent" as well as lucid. It might be suggested that the *combination* of lucidity and coherence sets near-death experiences apart.

We grant that this combination is important and distinctive of near-death experiences. We are even willing to grant, further, that it gives *some reason* to believe that their contents are accurate. But this is not all that is being claimed. The quotations from Alexander and Long suggest that the vividness and coherence of near-death

9. Sacks (2012a: 39).
10. Sacks (2012a: 229).

experiences provide us with *overwhelming* reasons to believe in the accuracy of these experiences. They claim that we simply cannot doubt their accuracy. This is hasty. Near-death experiences may be distinctive in combining vividness and coherence, and are thus importantly different from dreams and hallucinations. But it does not follow that they give us *overwhelming*, or even *strong* evidence for supernaturalism. At most, this special combination of vividness and coherence provides *some reason* to take supernaturalism seriously. It certainly does not tip the scale in its favor all by itself. The case for supernaturalism should be cumulative, built on various pieces of independently plausible evidence. If the vividness and coherence of near-death experiences are to convince us of supernaturalism, they need some help.

This last point may seem trifling. Why harp on the inconclusiveness of the line of argument that begins from the special vividness and coherence of near-death experiences? So what if some, such as Alexander and Long, have overstated the case? The important thing is to take the evidence for supernaturalism seriously. And they are calling our attention to the need to do so.

We agree that the evidence for supernaturalism should get a fair hearing. Indeed, that is one aim of this book. But just as it would be a mistake to dismiss supernaturalism out of hand, it would also be a mistake to lose track of sound principles of reasoning. To use an analogy with a legal trial, we are not supposing that the defense attorney for physicalism is in a position to demand that the judge dismiss the case! Rather, our point is that (thus far at least) the jury should not be persuaded that a clear and convincing case has been made for supernaturalism.

Many of the theorists who have weighed in on the meaning of near-death experiences are physicians or scientists, such as Eben Alexander, Jeffrey Long, and Pim van Lommel. And they have sometimes asserted that as physicians or scientists they have a kind of special authority in this intellectual territory. For example, in personal correspondence with us, the authors, Pim van Lommel wrote:

> ... I have some questions. ...
>
> First: What is the profession of both you [John Martin Fischer] and of Ben Mitchell-Yellin?
>
> Second: How many people with an NDE did you interview, and in how many of these patients did you personally corroborate veridical perception that must have happened during their NDE?
>
> Third: Did you ever attend a patient during cardiac arrest with CPR, and what was your impression about the clinical state of those patients?[11]

Here van Lommel is clearly questioning our competence, as philosophers, to enter into productive analysis and discussion of near-death experiences. He is also conflating the two senses in which near-death experiences may be "real." He asks whether we have interviewed patients to corroborate the veridicality of their experiences. But it is simply unclear how an interview with someone who reports having actually had an experience can shed light on whether that experience is accurate. These are two different issues.

We are willing simply to grant that those reporting near-death experiences actually had the experiences they report having. That is, we are willing to grant the reality of these experiences in the subjective sense of the term "real." But this should not be confused

---

11. Email correspondence (7 January 2014).

with the assumption that the content of these experiences corresponds to reality. In some cases it may take work to establish that an experience one actually has represents the world as it actually is. Near-death experiences are a special case in this regard. Not all experiences are like this. We are not thoroughgoing skeptics, withholding our assent to the claim that any experience faithfully represents reality. We are not of the opinion that we are living in the Matrix (a phony world that does not reflect true reality, as in the film by that name). But near-death experiences are not normal experiences. Indeed, this is why they are of such great interest. We want very much to know whether they are putting those who have them in touch with reality, and if so, how.

In this sort of analysis, a physician has no special claim to authority. Indeed, we do not think that it takes a PhD in philosophy (or any other discipline) to recognize the difference between really having an experience and the accuracy or truth of what that experience represents. Thoughtful people, no matter their profession, can see the space for doubt about vivid and lucid experiences. It is not as if the box office success of *The Matrix* was due to the moviegoing habits of academics alone! It takes no special training to be able to reflect carefully and seriously about the implications of near-death experiences for issues about the reality of the external world and our access to it.

In the above correspondence, van Lommel was responding to an article we published in reply to his in the *Journal of Consciousness Studies*. In that article, we expressed the thought that people who have had near-death experiences, and people in general, would want to accept certain implications of these experiences (such as the notions that the mind is nonphysical and that there is an afterlife) "with their eyes wide open." They would want to do so with a clear understanding of alternative perspectives. An anonymous reviewer for the journal wrote in his or her comments to the journal editor and to us:

> I wonder how many NDErs the authors have actually spoken with. They may or may not be interested to know that the vast majority of NDErs believe that it is the authors whose eyes have not yet open[ed]. But they hold out hope.

But, again, we are perfectly willing simply to grant that many people really do have near-death experiences. We do not think that they are making these experiences up. And we are not doubting that the experiences these people report having really do contain what they tell us that they contain—for example, trips to otherworldly realms, conversations with deceased loved ones. Perhaps speaking with subjects of near-death experiences would help a skeptic to fully acknowledge the reality, in this weaker sense, of the experiences. Perhaps it would squash skepticism about the sincerity of people's reports about their near-death experiences.

But the fact is that we are not skeptical about the sincerity of people's reports. We are not calling into question the reality of near-death experiences in this sense. This does not get at the heart of the matter. The idea that speaking with people who have had near-death experiences would help someone to recognize that these experiences are *accurate and verify the existence of the afterlife* just doesn't make sense. We cannot see how speaking with someone who has had a profound experience involving a trip to heaven or who has witnessed his own body undergoing CPR could possibly show that the content of that experience was accurate.

All manner of human experiences seem to represent reality when they are actually illusory. Just to use mundane examples: a straight stick might appear bent when viewed while it was stuck in water; an airplane might seem to be traveling slowly, if we see it from a certain angle; and so forth. If we may put it this way, illusions are for real! And we are not contesting the reality of near-death experiences in this sense of "real." We are claiming, however, that simply speaking to someone who has had a near-death experience would be insufficient to show that this kind of experience is fundamentally different and beyond the shadow of doubt when it comes to the accuracy and truth of what it represents. What van Lommel and others are prescribing—for us to just go out and talk to people who have had near-death experiences and we will be the ones who see the light—would be useless in assessing whether these experiences match external reality.[12]

Consider religious belief. When you go to churches, synagogues, mosques, and other places of worship the world over and talk to the people there, many of them express their religious convictions with great emotion and certainty. And there is no reason to doubt that many of these good folks are entirely, sincerely, and deeply committed to the tenets of their religious faiths. Their conviction is impressive; it helps grab one's attention. You listen to people deeply convinced about what they are telling you. By contrast, you ignore slick preachers and snake-oil salesmen obviously out for your money. You brush them off as having ulterior motives. This is as it should be. But sincerity and conviction, even of the deepest kind, are not, in themselves, evidence of truth. Particular religious beliefs may in fact be true, but the grounds for taking them to be true are not to be found in the sincerity and earnestness

---

12. In his important and pioneering book, *Life after Life*, Raymond Moody (1975) emphasized the importance of personal contact and interviews with people who have had near-death experiences. But it is interesting to note that Moody falls short of affirming supernaturalism.

of these believers. They are to be found in the contents of what these believers tell us. We can put the point this way: sincerity grabs the ear, but only good evidence compels belief.

Many people, and not just those who have had near-death experiences, hold out great hope for the afterlife. But, in our view, people want *genuine* hope. They want to believe in something that is both convincing and real. They want to believe for the right reasons, and not just because someone is telling them what they want to hear. People genuinely want to figure out how good their reasons are for their conclusions. We have been pointing out here that sincerity and conviction are not themselves reasons to believe anything. At most, they are *signals* that the reasons to follow are worth taking seriously. The merits of a conclusion depend, ultimately, on the evidence marshaled in support of it.

Chapter 6

Near-Death Experiences
in the Blind

There are various reasons to be skeptical of what other people tell us, especially if what they are telling us sounds farfetched. We might doubt their motives. Are they telling the truth or bilking us? If they have something to gain from convincing us of what they say, we would be wise to consider their claims with a grain of salt. For example, if their goal is to sell us something, we should carefully consider whether what they say is true or, rather, designed to get us to dip into our wallets. We might also doubt their standing to testify about what they are telling us, independent of their motives. If what they say sounds fanciful, we might begin to wonder if they even know what they are talking about.

As we mentioned in Chapter 1, those who report near-death experiences are often highly attuned both to the fact that they are asking their audience to believe something quite out of the ordinary and that this requires establishing their credentials on the matter. It is no accident that medical professionals are well represented in the ranks of authors on books about near-death experiences.[1] They are experts about our bodies and brains. We

---

1. For instance, four authors on whose work we have focused—Alexander, Greyson, Long, and van Lommel—are all MDs. Several of them make sure to remind us of this quite often.

respect our doctors and defer to their authority in weighty, intimate matters all the time. They have trained extensively to acquire the knowledge necessary to help us in very important ways. They help us stay alive and healthy. We should respect them for that. But while the question of what one should do to keep one's blood pressure down is one thing, the question of whether near-death experiences prove that there is a heaven is another. Doctors may be authorities on a great deal that matters to us. But that does not mean that they are especially well suited to tell us about the implications of near-death experiences.

It is also no accident that those who appeal to near-death experiences in order to argue for supernaturalism invoke the testimony of young children.[2] This also seems to make good sense. After all, whose motives could be purer than those of a young child? Who is better suited than a young child to relate his or her experience in a manner that is free from the intrusion of cultural baggage? It seems that the testimony of children ought to be especially free from suspicion. But those children who have near-death experiences are not the primary disseminators of the reports about these experiences. It takes an adult to put the words of a young child into print, pitch the manuscript to a publisher, sign the contracts, and so on. It takes an adult to lay out the implications of a young child's experience for the relationship between the mind and the body or the relationship between this world and the next. It would be hasty to conclude that the testimony you read coming from a young child is untainted by adult interpretation and motivations. Our usual reasons to be skeptical still apply.

2. We will discuss near-death experiences in children in Chapter 7, and we will consider in detail one such case in Chapter 11, when we discuss Burpo (2010).

There is, however, one particular population of people who have had near-death experiences with respect to whom the usual skepticism seems beyond the pale. Some blind people, even some who have been blind from birth, report having had near-death experiences that include visual representations. And this suggests a powerful line of thought in support of the conclusion that an adequate explanation of near-death experiences must appeal to the nonphysical. The following anecdote from Long illustrates this line of reasoning well:

> This takes me back to the second question that the research has posed to me, namely: What does it mean that a blind person can see during an NDE?
>
> I had never even thought of this as a question until I took my son Phillip to a meeting in Seattle where a blind woman talked about her NDE. Phillip was nine years old at the time, and I thought he would be bored with the presentation. But his response was quite the contrary. Her presentation held his rapt attention. When the lecture was over, we walked quietly to the car. I could tell something was on Phillip's mind, so I said nothing, inspiring him to fill the dead air.
>
> Finally, Phillip spoke. "If blind people can see during a near-death experience, then the experience must not be caused by brain chemistry," he said. "The experience must be real."
>
> It's thinking like this that puts blind sight high on my list of evidence for the afterlife.[3]

---

3. Long (2010: 91–92).

This is a compelling story. But it suggests several distinct lines of thought that need to be kept separate.

Let's begin by considering the sense in which these experiences are supposed to be real. In one sense of "real," we are happy to admit that these experiences are real. We do not doubt that some blind people really have had the visual experiences they claim to have had. But we need to distinguish between the claim that these experiences are real, in the sense that they really did happen to the person who had them, from the claim that they are real, in the sense that they truly and accurately represent the world. Dreams and hallucinations are real conscious experiences in the first sense, but not necessarily in the second. Similarly, a visual experience had by a blind person may be real in the one sense but not the other. She may really experience visual phenomena that do not accurately represent the world the way visual phenomena normally do for the sighted.

Moreover, we need to be careful in considering the relevance of the reality of an experience, in either of the above senses, to the question of whether it is the effect of physical causes. All experiences that are real in the first sense, all those that actually occur, call out for explanation. But it is not at all clear why the content of an experience, and, in particular, its accuracy, should make a difference with respect to what the terms of an adequate explanation would be. Why should the fact that an experience truthfully depicts reality suggest that it has a nonphysical explanation? Surely there is no tension between an experience being "caused by brain chemistry" and its being real in both senses at issue. To claim otherwise would be to claim that everyday perceptual experiences—seeing the tree outside one's window—must not have physical causes, because they actually occur and faithfully represent the world. That is, it would be absurd to claim that we must appeal to nonphysical mechanisms—something supernatural—in order to explain our really seeing a tree that is actually there. We doubt that anyone (in this discussion) is seriously making this claim.

The reasoning illustrated in the anecdote about Phillip, however, has to do with near-death experiences in the blind. And it may seem as though there is a real difference between cases of everyday perceptual experiences and visual experiences in the blind, especially if these visual experiences faithfully represent the world. How could a blind person have an accurate and true visual experience? And how could such an experience be explained by physical mechanisms?

Though near-death experiences in the blind might seem at first blush like convincing cases in support of supernaturalism, we think that it is plausible that there are physical explanations of these experiences. Begin by distinguishing two kinds of blindness. In the first kind of blindness, the blind individual is unable to receive visual information as an input to his or her brain from the outside world. The normal ways in which our eyes receive information that is then communicated to and processed in the brain to generate visual impressions are damaged and occluded. Call this "no-input blindness." The brain of one with no-input blindness might still be capable of generating visual impressions. Whatever neural mechanisms are involved in processing inputs to generate visual impressions may be able to function properly, even though the mechanisms that provide the material for such processing—the eyes and retinal nerves—do not function properly. (Perhaps they do not receive information at all, or perhaps they simply cannot communicate this information to the brain.) The situation of these people would be such that, were their brains to receive inputs from their eyes, they would be able to see (perhaps after a "learning" period). But they cannot actually see, because their brains do not receive input from their eyes.

In principle, at least, someone with no-input blindness could have visual impressions and experiences prompted by physical means other than the usual way of receiving input from her eyes. It might be possible, for instance, to hook someone's brain up to a mechanism, other than her eyes, from which her brain may receive the same kind of information that would normally be communicated by one's eyes. This would be like giving her prosthetic eyes. Because her brain is capable of processing this information in the normal way, there would seem to be no barrier to her having visual experiences. Indeed, there would appear to be no barrier to her having accurate visual experiences, given that the prosthetic eyes from which her brain is receiving inputs communicate the way things look in such a manner that her brain can generate visual impressions that represent reality as it actually is. And all of this seems perfectly explainable in physical terms. (As we will see in a moment, this "in principle" argument has a corollary in fact— there are blind people who experience visual dreams and hallucinations. And we can explain them in physical terms.)

So it seems that near-death experiences involving subjects with no-input blindness do not pose a significant threat to physicalism. But now consider a second kind of blindness. In this case, the mechanisms that receive and communicate visual information to the brain may function normally, but the mechanisms in the brain that process this information are nonfunctioning. An individual with this kind of blindness is unable to generate visual impressions because his brain is unable to process inputs. Call this "processing blindness." It does seem inexplicable how we might account for visual experiences had by an individual with processing blindness in physical terms. Even if it were possible to provide input of visual information to the brain of one with processing blindness, by the normal means or in some other way, it would not be possible for this person's brain to generate visual impressions from this information. The physical mechanisms in this person's brain are incapable of doing so. Thus, processing blindness is crucially different from no-input blindness.

A near-death experience had by one with processing blindness would provide a much more compelling case for the need to appeal to nonphysical mechanisms in order to make sense of the experience. The simple reason for this is there does not seem to be any way of explaining the manifestation of a visual experience in someone with processing blindness in physical terms. It does not matter if we can explain, in physical terms or in any other way, the acquisition of information that might form the basis for a visual experience. By definition, processing blindness involves impaired functioning of the physical means by which visual representations are constructed. Thus, it would seem to be a great mystery, from the physicalist's perspective, how someone with processing blindness could come to have a visual experience of any kind. It would be a mystery if such a person had a visual experience that did not correspond to reality, and it would be an even greater mystery if he or she had one that did correspond to reality.

A blind individual might have both kinds of blindness or only one kind. If we are correct that the kind of blindness one has matters in the way we have just described, then it makes a difference whether those cases of near-death experiences in the blind are had by those with no-input blindness or with processing blindness. If we do not know which of these two kinds of blindness exists in a given case, then we cannot be certain whether it is a case that rules out a physical explanation of the experience. This provides us with good reason to be cautious in interpreting cases of near-death experiences in the blind. We cannot simply conclude from the fact that a blind individual had a near-death experience with visual content that this experience cannot be explained in physical terms. If it is a case of no-input blindness, as we have argued, it may well admit of an explanation in physical terms.

It turns out that there actually *are* many well-documented cases of visual hallucinations in the blind, especially as these occur in people afflicted with Charles Bonnet syndrome (CBS).[4] Here is a particularly interesting description of such a patient offered by Sacks:

One day late in November 2006, I got an emergency phone call from a nursing home where I work. One of the residents, Rosalie, a lady in her nineties, had suddenly started seeing things, having odd hallucinations which seemed overwhelmingly real. The nurses had called the psychiatrist in to see her, but they also wondered whether the problem might be something neurological. . . .

When I arrived and greeted her, I was surprised to realize that Rosalie was completely blind—the nurses had said nothing about this. Though she had not seen anything at all for several years, she was now "seeing" things, right in front of her.

"What sort of things?" I asked.

"People in Eastern dress!" she exclaimed. "In drapes, walking up and down stairs . . . a man who turns towards me and smiles, but he has huge teeth on one side of his mouth. Animals, too. I see this scene with a white building, and it is snowing—a soft snow, it is swirling. I see this horse (not a pretty horse, a drudgery horse) with a harness, dragging snow away . . . but it keeps switching. . . . I see a lot of children; they're walking up and down stairs. They wear bright colors—rose, blue—like Eastern dress." She had been seeing such scenes for several days.

. . . I explained to her that hallucinations, strangely, are not uncommon in those with blindness or impaired sight, and that these visions are not "psychiatric" but a reaction of the brain to the loss of eyesight. She had a condition called Charles Bonnet syndrome.[5]

Although the hallucinations subsided within a few days, Sacks writes:

Almost a year later, though, I got another phone call from the nurses, telling me that she was "in a terrible state." Rosalie's first words when she saw me were "All of a sudden, out of a clear blue sky, the Charles Bonnet has come back with a vengeance." She described how a few days before, "figures started to talk around, the room seemed to crowd up. The walls turned into large gates; hundreds of people started to pour in. The women were dolled up, had beautiful green hats, gold-trimmed furs, but the men were terrifying—big, menacing, disreputable, disheveled, their lips moving as if they were talking."

In that moment, the visions seemed absolutely real to Rosalie.[6]

5. Sacks (2012a: 3–4, 5).
6. Sacks (2012a: 8).

Indeed, Sacks points out that the hallucinations of Charles Bonnet syndrome patients seem not only absolutely real to them, but they are also often lucid and detailed: "CBS hallucinations are often described as having dazzling, intense color or a fineness and richness of detail far beyond anything one sees with the eyes."[7]

No one, however, takes seriously the notion that these very lucid, real experiences in the blind are anything other than hallucinations entirely explained in terms of physical phenomena. In the last two decades, there has been considerable progress in understanding the neural basis of visual hallucinations.[8] This work appears to provide promising resources to explain what is occurring in the cases of those with Charles Bonnet syndrome. And this should raise our confidence in the prospects of explaining the causes of visual representations in near-death experiences had by the blind. Even if these two kinds of visual experience in the blind are not caused via the same physical mechanisms, our understanding of what is happening in the one case is likely to illuminate what is happening in the other. And, importantly, the fact that there are plausible physical explanations of visual experiences in patients with Charles Bonnet syndrome suggests that explaining visual representations in near-death experiences reported by blind people is not as daunting a task as it may at first seem.

It is interesting to note that just as there can be visual hallucinations among the blind, there can be musical hallucinations among the deaf. Sacks writes:

Although musical phrases or songs may be heard along with voices or other noises, a great many people hear only music

---

7. Sacks (2012a: 22).
8. See Ffytche (2007). See also Ffytche et al. (1998). Ffytche's work (including his collaborative work) is discussed in Sacks (2012a: 24–27).

or musical phrases. Musical hallucinations may arise from a stroke, a tumor, an aneurysm, an infectious disease, a neurodegenerative process, or toxic or metabolic disturbances. Hallucinations in such situations usually disappear as soon as the provocative cause is treated or subsides.

Sometimes it is difficult to pinpoint a particular cause for musical hallucinations, but in the predominantly geriatric population I work with, by far the commonest cause of musical hallucination is hearing loss or deafness—and here the hallucinations may be stubbornly persistent, even if the hearing is improved by hearing aids or cochlear implants.[9]

Again, just as with visual hallucinations in the blind, no one thinks that the auditory hallucinations in the deaf are anything other than hallucinations that are in principle explainable in physical terms.

Indeed, Sacks points out that "the musical hallucinations of deafness and the visual hallucinations of CBS may be akin physiologically."[10] He notes that visual hallucinations involve many of the same neural mechanisms as ordinary visual perception, just as musical hallucinations involve many of the neural mechanisms of actual auditory perception. There is simply no need to posit nonphysical mechanisms in either case.

It might seem that there is a tension here between taking the expert word of a neurologist like Oliver Sacks and our claim in Chapter 5 that "expert medical knowledge" is not needed to make meaningful interpretations of near-death experiences. It might seem to some that there is a double standard here—we make use of scientific credentials when they support our

9. Sacks (2012a: 66).
10. Sacks (2012a: 73).

argument, and discount them when they do not. But this charge relies on ignoring an important distinction between relying on expertise *when it comes to the subject matter of that expertise* and relying on expertise *even in matters that go beyond that subject matter*. We certainly *do* think that medical scientists are in a good position to study and understand the workings of our brains and perceptual systems. A neurologist is clearly authoritative about the neural mechanisms of ordinary visual perception and so has some authority about their relationship to the mechanisms of various kinds of hallucinations. A cardiologist is an authority on the cardiovascular system, and so also on its relationship to brain functioning. No debate here. What we denied in Chapter 5 was that simply by talking to patients who have had near-death experiences or even carefully cataloguing their responses, a medical doctor is in a special position to assess the meanings of these experiences. The meaning of experience is not the subject matter of medical expertise. And this is why it is reasonable to appeal to authority in the present case but deny it in the previous one.

To sum up, we have seen several reasons for being cautious in drawing conclusions about the need to appeal to nonphysical mechanisms in order to explain near-death experiences in the blind. The first reason for caution is that there may be different causes of blindness, not all on a par with respect to the challenges they pose for finding an adequate physical explanation of a near-death experience with visual content. The second reason is that there are well-documented instances of visual experience in the blind, such as those who have Charles Bonnet syndrome, and these other instances are not taken to pose a challenge for physical explanations. We should look before we leap to supernaturalist conclusions, like that of Long and his son, Phillip.

There is an obvious reply to be made on behalf of those who take near-death experiences had by the blind to support supernaturalism. Even if it would be possible to provide physical explanations of *some* cases in which blind people have near-death experiences involving visual phenomena, it may not be possible to provide physical explanations of *all* or even *most* such cases. The distinction between processing blindness and no-input blindness is only helpful for the physicalist if all or most of those blind people who have near-death experiences involving visual phenomena are people with no-input blindness. As we have already admitted, there does not seem to be a ready physical explanation of such experiences had by those with processing blindness.

This has the potential to be a very forceful challenge. But we do not know exactly how forceful because we do not know whether the cases cited by Long and others in support of supernaturalism are cases of people with processing blindness or no-input blindness. The distinction is simply not on the radar. It would be helpful if it were. However, there is reason to think that if we were to look into the causes of blindness in those people who have near-death experiences (and we think this would be an excellent research program!), we would be likely to find that these are cases of no-input blindness.

Though there are many causes of blindness, the most prevalent ones (e.g., glaucoma) involve conditions in the eye. To be fair, we should note that there are also cases of blindness caused by conditions in the brain (e.g., cerebral visual impairment). These are cases of processing blindness. But processing blindness is rarer than no-input blindness.[11] Hence, while we cannot rule out the possibility

11. For some overviews of the different causes of blindness, see World Health Organization (2014), National Institutes of Health (2008), Blaikie (2014), and the American Foundation for the Blind (2014).

that there are some with processing blindness who have near-death experiences involving visual phenomena, we also cannot rule out the possibility that absolutely every blind person who has a near-death experience involving visual phenomena has no-input blindness. And if we were forced to bet on which kind of blindness is implicated in a case of a blind person who has had a near-death experience, no-input blindness would be the safe bet, especially because it is more common. The two possibilities are not on a par. The upshot is that, given our current state of knowledge about these matters, it does not seem as though near-death experiences in the blind provide a very forceful challenge to physicalism.[12]

12. For further helpful discussion of near-death experiences in the blind, see Blackmore (1993).

Chapter 7

Near-Death Experiences in Children and throughout the World

The fact that near-death experiences the world over are very similar seems to support the conclusion that these experiences require nonphysical explanations. If children and adults all over the world are reporting near-death experiences with similar content, then it seems that this content is not influenced by age, location, culture, or other contingent factors having to do with the particularities of one's life or physical circumstances.[1] This appears to provide good reason to believe that near-death experiences put one into contact with a unified, objective, and nonphysical reality. How else could such disparate individuals come to have such similar experiences? In this chapter we explain why we think this reasoning is faulty, as is similar reasoning based on near-death experiences involving life reviews (components of NDEs in which "highlights" of one's life appear) and near-death experiences involving reunions with loved ones.

1. For a classic study of near-death experiences in children, see Morse (1991).

In the following passages, Long clearly presents the line of thought we wish to argue against:

> The near-death experiences of children, including very young children, are strikingly similar to those of older children and adults.[2]

> *The core NDE experience is the same all over the world*: Whether it's a near-death experience of a Hindu in India, a Muslim in Egypt, or a Christian in the United States, the same core elements are present in all, including out-of-body experience, tunnel experience, feelings of peace, beings of light, a life review, reluctance to return, and transformation after the NDE. In short, the experience of dying appears similar among all humans, no matter where they live.[3]

> This is further evidence that NDEs are much more than simply a product of cultural beliefs or prior life experiences. Near-death experiences remind us that although the people on earth may be a world apart, they may share this important spiritual experience. It's amazing to think that no matter what country we call home, perhaps our real home is in the wondrous unearthly realms consistently described by NDErs around the world.[4]

Long invokes the consistencies between near-death experience reports in adults and children and the consistencies in reports from around the world as proof that these reports are not culturally conditioned. Rather, he claims, they are reports of genuine

2. Long (2010: 200).
3. Long (2010: 150).
4. Long (2010: 171).

connections with "unearthly realms." This is certainly *one* possible explanation of the consistencies. If everyone—children and adults throughout the world—were in contact with the same nonphysical realm, then we would expect their reports of their near-death experiences to be similar, at least on the assumption that their access to this unified realm is reliable and accurate.

But this is not the only possible, nor the most plausible explanation of the phenomenon of worldwide consistency. Long looks to a similarity in the putative *objects* of experience, but it is at least as natural and plausible to consider a similarity in the *experiencers*. Human beings throughout the world have similar brains and central nervous systems. And although the brains of children are not fully developed, they also share basic similarities in biological makeup and function with adult human brains. Given these fundamental physical similarities in those who have near-death experiences at different ages and places the world over, one would expect consistencies in their experiences in similar circumstances. We should expect that those of similar physical constitution in similar circumstances, namely, near-death contexts, to have similar experiences. We do not need to look beyond our fundamental biological similarity for an explanation of the similarities in near-death experiences of children and adults the world over.

This is not to deny that cultural context may play a role in shaping the reported contents of near-death experiences. Surely our social milieu shapes the way we experience things, and we would need a compelling reason to suppose that near-death experiences are different in this regard. Perhaps the supernaturalist would like to deny that cultural differences manifest in dissimilar reports of near-death experiences. But this would seem hasty. Even if near-death experiences put people in touch with a single, unified supernatural realm accessible to all, it would still be eminently plausible that people from different cultures would interpret and report this realm in different ways. The same goes for our claim about biological similarity. Even if the similarity in near-death experiences the world over is to be explained by the similarity in the way human beings are put together the world over, this does not preclude the possibility that people's reports of their experiences are culturally conditioned. Neither side in our present debate gains the upper hand if reports of near-death experiences prove to be sensitive to cultural context.

We are not arguing that the evidence from near-death experiences is absolutely incompatible with the hypothesis of an "unearthly realm" (or with the hypothesis of nonphysical mechanisms of experience). Surely, given the evidence, it is *possible*, that people who have near-death experiences thereby come into contact with a single nonphysical realm. This is one possible explanation of the similarities in near-death experiences. But the *mere possibility* of a nonphysical realm is not what is at stake in discussions of the implications of near-death experiences. Many of those writing about near-death experiences claim that the evidence from them renders supernaturalism *highly probable*, or even claim that we are in some sense required to accept supernaturalism in order to make sense of these experiences. This is what we reject. It would seem *at least as plausible* (if not more plausible) that the physical similarities in human experiencers explain the consistencies in near-death experiences as that this consistency is explained by contact with a nonphysical realm. We do not take issue with the mere possibility of a nonphysical explanation but rather with its relative plausibility in comparison with rival physical explanations.

Even if the fact of similarities in near-death experiences the world over does not support the claim that these experiences reveal the reality of a nonphysical realm, there may be other, more promising avenues to this same conclusion. Many, though not all, people who have near-death experiences have a component that is described as a "life review"—a kind of "highlight reel" of one's life. There are many interesting questions about the nature and significance of this component of near-death experiences.[5] One of these is whether the life review provides support for supernaturalism.

Once again, a quotation from Long illustrates this line of thinking. "Accurate and transformative life reviews are a hallmark of NDEs, and they point to a reality beyond what we know from our earthly existence. They provide important evidence for the reality of an afterlife."[6] The move here is from the claim that many near-death experiences contain an accurate review of the subject's life, one that represents actual events from the subject's past experience, to the claim that this supports the reality of an afterlife. We grant that life reviews are often genuine parts of near-death experiences and that they are frequently or typically accurate. We do not dispute the claim that these experiences are "real" in both of the senses previously discussed—people actually have near-death experiences with life-review components and their contents correspond to actual events in their personal histories. But why suppose that these features of life reviews "provide important evidence for the reality of an afterlife"? This just does not follow. It is not clear that there is any connection at all between reviewing past experiences in this life and gaining insight into the hereafter.

The life-review component of many near-death experiences does not support the first tenet of supernaturalism. Nor does it support the second tenet of that view. That is, this aspect of near-death

5. Some particularly fascinating work is reported in Katz, Grosman-Saadon, and Arzy (ms.).
6. Long (2010: 120).

experiences does not point to the conclusion that these experiences cannot be explained in physical terms. To see this, consider two different ways in which one might be said to remember something. Although many of the events featured in a life review may have been forgotten in a colloquial sense, it does not follow that they are not still encoded in the brain. Memories of these events may be stored in the brain but unavailable for retrieval in the normal way. There is a sense in which they are forgotten, since the individual cannot recall these memories as she can other things she is said to remember. But there is also a sense in which the individual remembers these events. They are stored in her brain and may be recalled, just not in the normal way. Whatever else a near-death experience may reveal, it is possible that it reveals memories of events that are occluded from one's normal powers of recall. And this possibility can be made sense of in wholly physical terms.

Psychologists refer to the kinds of memories we are positing here as "implicit memories." Implicit memories are unconscious. They may influence one's behavior, but they are not available to the gaze of conscious reflection. By contrast, "explicit memories" are consciously recollected. These are what we normally refer to as memories.[7] Philosophers also contend that not all memories must be available to consciousness, and this contention does not rely on denying physicalism.

Although we do not yet know the exact physical mechanisms of storage and "playback" of life reviews, the mere fact that these experiences involve the recall of events that are not part of one's

---

7. For an interesting overview of implicit memory, see Schacter (1987). And see especially, on p. 502 of Schacter (1987), the striking discussion of Descartes' comments in *The Passions of the Soul*, where he says that "a frightening or aversive childhood experience may 'remain imprinted on his [the child's] brain to the end of his life' without 'any memory remaining of it afterwards.'"

conscious memory bank does not indicate the falsity of physicalism. The fact that we do not yet have an adequate physical explanation for the life-review component of near-death experiences does not indicate that no such explanation is forthcoming. It merely puts pressure on us to seek a plausible physical explanation. Our scientific understanding will continue to improve. We should not take our current lack of understanding regarding a particular phenomenon to be cause for despair about the possibility of arriving at an adequate explanation of it in the future. This would be a radically anti-scientific stance. The scientific enterprise is fueled by our quest for making sense of things we currently do not fully comprehend in terms of the measurable and observable. In other words, it is our quest for physical explanations that drives scientific inquiry. To hold that lack of an adequate physical explanation of something is grounds for rejecting the possibility of any such explanation is to give up on this pursuit of understanding. And there is no good reason to do that.

Again, it is puzzling to claim that the existence and intensity of life reviews would indicate the presence of an afterlife. After all, the events that are supposed to be recalled during these life reviews originate from *within* the individual's life. These events do not reach beyond the world as we normally know it. They do not pertain to an afterlife, or even to a nonphysical realm. Thus, even if it is as yet unclear how exactly the life reviews are stored in the brain and triggered in near-death episodes, it is a stretch to posit an afterlife as a way of explaining them.

There is, however, another aspect common to many near-death experiences that appears to provide a more convincing basis for the conclusion that they put one into contact with a nonphysical realm. Many people who have had near-death experiences report being reunited with loved ones, and they report experiencing the reunion as taking place in an otherworldly realm.[8] It may seem as though the widely shared experience of seeing and communicating with the deceased while undergoing a near-death experience is evidence that these experiences put one in touch with a nonphysical reality. After all, these are dead people who no longer exist in the physical world around us. It seems as though they must exist in nonphysical form somewhere else in order to be active participants in one's experience.

The first question to ask about this aspect of near-death experiences is why an individual would be reunited with *his or her* deceased relatives, as opposed to strangers. The question remains pressing even if we stipulate that there is an afterlife. Consider heaven. Supposing that heaven exists and an individual in a near-death episode is transported there, how exactly is he or she connected specifically to his or her own relatives in heaven? Even with the assumption of an afterlife, the specific means whereby an individual is connected to his or her own relatives, as opposed to the many other people in heaven, is left for further explanation. In other words, the simple postulation of an afterlife does not do the explanatory work we need.

We want to know why people who have near-death experiences come into contact with *their* loved ones. Simply pointing to the reality of an afterlife does not provide an adequate answer. At most, it provides the backdrop for a certain kind of more detailed answer. For example, it provides some of the material out of which to forge an explanation that posits some special force connecting the souls of particular deceased individuals—the subject's loved

8. Among the near-death experiences discussed at length in this book, this feature is a part of the ones experienced by Pam Reynolds, Eben Alexander, and Colton Burpo, whose near-death experience we discuss at length in Chapter 11.

ones—with the subject of the near-death experience. But it does not explain the force behind this connection, nor how it is supposed to produce the reunion. Perhaps the assumption is that the force is love and the connection is guided by the hand of God. But, we wish to stress, these claims go well beyond the claim that the afterlife is real, and it is not at all clear how the relevant contents of near-death experiences are supposed to establish these further claims.

We shouldn't need to appeal to elaborate, nonphysical explanations of this sort in order to make sense of this aspect of near-death experiences. Terror management theory offers a plausible, physicalist-friendly explanation of this phenomenon.[9] It proposes that human beings have special capacities, such as self-awareness and abstract thought, which allow us to be aware of our own mortality and to ponder the potential absurdity of the world around us. We can entertain the thought that there is no good reason why we are here at all. And this can be terrifying; so can the thought that our lives are destined to be extinguished. We can imagine a world without us. Much of what we recognize as human culture, according to this theory, has arisen as a means of dispelling the terror we bring on ourselves by thinking these all-too-human thoughts. In order to defend against our fear, we buy into value systems and cultural practices that we take ourselves to be living up to and advancing. We combat our existential terror by building self-esteem.

Terror management theory is a very well-validated psychological theory. Over the past three decades, its predictions about

9. For a seminal discussion of terror management theory, see Greenberg et al. (1986). For a recent meta-analysis of the mortality salience hypothesis central to terror management theory, see Burke et al. (2010). For a comprehensive overview of the theory, see Solomon, Greenberg, and Pyszczynski (2015). Terror management theory is founded on the insights of Ernest Becker (see, especially, Becker (1973)).

human behavior have been repeatedly verified. Important for our purposes here, all of the key elements of terror management theory are compatible with a physicalist theory of mind and the firm conviction that there is no afterlife. Thus, any explanation of the presence of deceased relatives in near-death experiences by appeal to terror management theory would be compatible with physicalism and the lack of an afterlife.

Terror management theory is certainly a reasonable view, and one that can provide a plausible explanation of the relevant feature of near-death experiences. A proponent of terror management theory might point out that we have various psychological mechanisms that give us comfort in the face of danger and fear. One such mechanism might well issue in visions of deceased relatives in near-death experiences. After all, the presence of such figures would presumably provide great comfort in the face of death. Reassurance that we will be reunited with our lost loved ones and that they will welcome us to the next phase of our existence would bolster our sense of living up to the ideals of being a good son or daughter, niece or nephew. It would enhance our self-esteem. The mere existence of strangers or people with whom we do not have close bonds—and family is perhaps the closest bond for most people—would not provide this same comfort. So terror management theory can explain why people who have near-death experiences would experience reunions specifically with *their* loved ones and not strangers, without postulating the reality of an afterlife or appealing to nonphysical mechanisms.

We contend that terror management theory provides a very plausible and physicalist-friendly explanation of the presence of loved ones in many near-death experiences. But it may strike some as importantly incomplete. Many people who have had near-death experiences report experiencing encounters with deceased relatives *whom they did not even know existed*. It is difficult to see how this may be explained in terms of a drive to comfort oneself in the face of death.[10] The face of someone unknown to one would seem relatively unremarkable, especially in a context awash with the faces of people one recognizes as loved ones. Thus, it is difficult to understand how the presence of this individual, an apparent stranger, could be a tool deployed in one's own battle against anxiety at the thought of one's death.

This may be so, but it does not pose an insuperable challenge to the adequacy of the terror management theory–based explanation we have offered. We do not think that the plausibility of the terror management approach depends on the subject of the near-death experience recognizing the person in question as connected to him *during* the near-death experience. It is enough that the presence of this individual in the subject's *subsequent report* of what he or she experienced can be explained by appeal to terror management theory. And it is possible that the subject's subsequent report of what he or she saw during the near-death experience, albeit completely sincere, does not match what he or she really did see during that experience. This mismatch between what one saw and what one later believes that one saw might be explicable by appeal to psychological principles similar to those invoked by terror management theory.

Consider, for example, Eben Alexander's report of having seen a sister he never knew he had during his near-death experience. He did not realize that this person was a relative of his during his near-death experience. It was not until much later that he made

10. For a statement of this objection, see Long (2010: 133). The phenomenon in question is present in the near-death experiences of Eben Alexander and Colton Burpo, both of whom had the experience of meeting a sister they did not know they had.

the connection.[11] This makes room for the possibility that he interpreted back into his near-death experience something that was not originally a part of it. And it may be possible to explain this in terms of psychological processes similar to those that feature in terror management theory. Alexander reports being confused by the presence of this individual in his experience and wondering why she was there. He wanted very much to understand this part of his experience. Perhaps he came to make the connection between his sister and this woman in order to dispel his confusion. An explanation of this sort appeals to something like a basic aversion to the psychological conflict that comes with not knowing why something happened. This is similar to a psychological tendency to experience existential anxiety at the thought that the universe may be meaningless and absurd. Terror management theory invokes the latter tendency to explain our motivation to create systems of value that allow us to imbue the universe with meaning. Perhaps the former tendency explains our tendency to confabulate and enhance details when telling of what has happened in our lives in order for the overall narrative to be more coherent.[12]

The combination of this core tenet of terror management theory and the posited aversion to confounded understanding points to a powerful challenge to supernaturalism. Both tendencies are consistent with physicalism. And both offer plausible explanations of phenomena that, according to the supernaturalist, were

---

11. It is not clear how much later this revelation came to him. In a recent *Esquire* article about Alexander, Luke Dittrich (2014) suggests a different version of events from the one Alexander presents in his book. As Dittrich tells it, it was not until about four months after his near-death experience, and after he had read several books about near-death experiences, including one with a story about a girl who met a deceased brother she did not know about, that Alexander looks at a picture that had been sent to him of the sister he never knew and makes the connection.
12. Bortolotti (2010) cites support for the claim that we possess this tendency.

supposed to outstrip the explanatory resources of our physicalist framework. Thus, these tendencies work in tandem to undercut the supernaturalist's challenge to physicalism on the basis of near-death experiences. And the response they offer on behalf of physicalism is all the more powerful because these two tendencies may be seen as related in an important respect. They are means of aiding our comprehension, in narrative terms, of the world around us.

It is clear that supernaturalists take near-death experiences to challenge the adequacy of scientific inquiry to quench our thirst for understanding. We want more than science can give us. The implication is supposed to be that we need to supplement science with stories, which are not observable and testable. Some of these stories will appeal to nonphysical phenomena, such as our fundamentally nonphysical nature and a fundamental nonphysical reality. They tell of souls and heaven. And they are invoked as a basis for accepting the limitations of a merely physical set of explanations. But this is to overreach. A psychological theory, such as terror management theory, can help us to make sense, in physical terms, of our need for narratives and of the specific elements of the narratives we actually employ. It can explain our need for stories to support the value systems we see ourselves as living up to. No challenge to physicalism follows simply from the fact that we human beings seek to make sense of the world and ourselves through both stories and science.

In the previous several chapters, we have resisted various tempting ways of reaching the conclusion that supernaturalism is true on the basis of various features of near-death experiences. We will now switch from playing defense to going on offense. In Chapter 8, we will argue that appealing to the nonphysical makes no explanatory progress. Given the project of understanding near-death experiences, there is no good reason to transcend the physicalist paradigm that gives the context for everyday explanations of what occurs in our lives. After that, we will consider the form of explanation we find most promising in trying to explain near-death experiences in physical terms. We will conclude by considering the prospects of our preferred, physicalist view in accounting for near-death experiences as deeply meaningful and transformative.

Chapter 8

Can Near-Death Experiences Be Explained by a Single Factor?

The scientific worldview is a natural outgrowth of human curiosity. Our quest for understanding has led us to adopt a point of view that satisfies that curiosity. We have developed reliable methods for attaining ever more knowledge and theories to synthesize it. Science conceptualizes the world in terms of observable entities and regular relationships, which allow us to formulate hypotheses about the way things work. This is not to paint science as a static enterprise. One characteristic of science, as we have pointed out, is its openness to change. New data and new theories replace the old at an almost dizzying pace. We come to accept that things were not as they seemed before. Sometimes the only way to make sense of our observations is to suggest new entities and relations or propose new theories about how things fit together. But science marches on, not skipping a beat.

There are, however, some phenomena that seem to challenge the whole enterprise. They threaten to rock the house that science built down to its foundations. Part of what is so fascinating about near-death experiences is that they confound our attempts at scientific understanding. We want to be able to explain why certain individuals have the experiences they do, and we want to be able to do so in familiar terms. But scientific explanations do not seem up to the task. Some notice the apparent difficulty of explaining near-death experiences in physicalist terms and conclude that we ought to widen our explanatory net, so to speak. Near-death experiences, they claim, show that an underlying assumption of the scientific worldview is simply false. It will not do to conceive of the world in terms only of observable entities and relations. We must accept the existence of the nonphysical as well as the physical in order to grasp the way things work.

We remain unconvinced. If we look closely at how proponents of this argument press their case, we can see that they rely on an unduly restrictive conception of what a good explanation must be like. Various theorists have pointed out that we do not as yet have a fully adequate explanation of near-death experiences in terms of any *single* physical factor. They then infer that it is *highly probable* that physicalism is false because it cannot adequately explain the phenomena associated with near-death experiences. But this inference relies on the implicit principle that adequate explanations must appeal to a single factor only. Why?

Lest you think we are just making this stuff up, consider these arguments from the literature on near-death experiences that involve this mode of reasoning.[1] Pim van Lommel discusses four candidate explanations of near-death experiences, each of which invokes a different physical factor: lack of oxygen, high carbon dioxide, oxygen deficiency, and the use of certain drugs.[2] In each case, his objection is that the cited physical factor fails as a

---

1. See Mitchell-Yellin and Fischer (2014) for previous discussion of these issues in relation to van Lommel (2013).
2. Van Lommel (2013: 22–24).

complete explanation of near-death experiences. He objects that certain explanations fail to explain *all cases* of near-death experiences. For example, he objects to appeals to lack of oxygen because near-death experiences can occur in instances where this is not an issue. Additionally, he objects that certain explanations do not explain *all features* of NDEs. For example, drug-induced experiences lack elements commonly found in near-death experiences, such as a life review.

We do not dispute van Lommel's contention that in each case these factors fail adequately to explain the relevant phenomena. Rather, we claim that he is implicitly invoking a certain standard for adequate explanations that ought to be rejected. First, van Lommel assumes that an adequate explanation must be able to make sense of *all* features of *every* near-death experience. He assumes that an adequate explanation must cast a very wide net, leaving no near-death experience and no feature unexplained. Second, van Lommel assumes that an adequate explanation must appeal to a single factor *only*. This is to deny the merits of an explanation that combines several physical factors at once. Rather than consider the possibility of a multi-factor explanation of near-death experiences, van Lommel simply assumes that a physical factor appealed to as an explanation of near-death experiences must explain every aspect of all near-death experiences *by itself*. Because none of the physical factors he considers satisfies these two criteria, he rejects each one as at all relevant to the explanation of near-death experiences.

A similar set of assumptions is evident in Eben Alexander's discussion. He systematically considers, and rejects, nine single-factor physical explanations of near-death experiences: (1) a "primitive brainstem program to ease terminal pain and suffering"; (2) the distorted recall of memories from deeper parts of the limbic system; (3) endogenous glutamate blockade with excitotoxicity (mimicking ketamine); (4) Dimethyltryptamine (DMT); (5) isolated preservation of cortical regions; (6) unusually high levels of activity among "excitatory neuronal networks"; (7) involvement of subcortical structures, such as the thalamus, basal ganglia, and brainstem; (8) a "reboot phenomenon" (a "random dump of bizarre disjointed memories due to old memories in the damaged neocortex"); and (9) "unusual memory generation through an archaic visual pathway through the midbrain."[3] That's quite a list! And yet, just like van Lommel, Alexander rejects each and every one of these factors as relevant to the explanation of near-death experiences on the grounds that none of them *by itself* explains all features of every near-death experience.

It is worth pausing here. It might seem surprising—even startling—that people who have thought long and hard about the issues we are discussing in this book, as van Lommel and Alexander have, would accept such an over-simplified and implausible picture of explanation. Presumably, upon reflection, they too would come to see the need to reject it. But the logic of their explicit writings clearly presupposes the objectionable explanatory principle we have identified. Sometimes people actually use principles that they would, upon more careful reflection, reject. Careful consideration of the logic of their argumentation shows that they (and we) should reject it.

We grant that van Lommel and Alexander are correct that none of these physical factors *by itself* suffices to explain all aspects of all near-death experiences. We do not dispute this claim. Nevertheless, we think that their rejection of all these possible physical factors as at least *relevant* to explaining near-death experiences rests on a

3. See Appendix B in Alexander (2012a: 185–188).

mistaken conception of what makes for an adequate explanation. In other words, we do not doubt their data but rather the logic they use, in combination with this data, to arrive at the conclusion that these factors are explanatorily irrelevant.

Alexander and van Lommel accept something like the following two general principles as criteria on adequate explanations.

Complete Explanation: Any complete explanation of near-death experiences must account for *all* aspects of *all* near-death experiences.

Single Explanation: Any adequate explanation of near-death experiences must invoke only *one* explanatory factor (perhaps against a background of "given" conditions).

The first principle, *Complete Explanation*, might seem uncontroversial. But accepting it signals an interest in a particular kind of explanation. Consider our everyday practice of seeking and providing explanations. One way of thinking about this practice is in terms of answering "Why" questions.[4] An adequate explanation is one that fills a gap in our understanding. When we ask the fire chief why the house caught fire, we begin with a shared background understanding of things: the house is made of flammable material, fires are started when flammable materials are set aflame, there was no one home, and so forth. An adequate explanation will allow us to make sense of what happened, given this background understanding. When we learn that a burner on the stove was left on, this further piece of information enables us to connect the dots. The flame on the stove somehow spread to other parts of the

---

4. For a very influential account of this view, see van Fraassen (1980).

kitchen and, eventually, throughout the house. Learning that the stove was left on plugs a gap in our understanding. In light of our background understanding, it answers the question "Why did the house catch fire?" and so adequately explains the blaze.

It should be obvious that not all house fires are explained in the same way. Some house fires are caused by unattended stoves while others are caused by lightning strikes or candles or cigarettes and so on. An adequate everyday explanation of a particular house fire would fail to satisfy *Complete Explanation* because it would fail to explain *all* house fires. So it seems that in accepting this principle one is looking for a specialized kind of explanation, not the kind we offer to each other in our everyday practice of explaining what happens. What we are normally interested in when we ask why something happened is just that: an explanation of why that particular thing happened at that particular time. We would not reject the fire chief's response if he told us that it was the unattended stove that started the house fire on the grounds that, just the other week, a house fire was started by a lightning strike. Perhaps *Complete Explanation* is best regarded as a principle that governs a specialized explanatory practice. In adopting this principle, we seek answers to specialized "Why" questions, given a shared body of specialized knowledge about the cause of fires.

Consider pyrologists, those engaged in scientific inquiry about fire. A pyrologist seeks to understand the underlying causes of fire at a different level of detail than the homeowner or neighbor. The pyrologist would not be satisfied to learn that the house fire was caused by an unattended stove. She would want to know why it is that the flame on the stove spread to the rest of the house. She would want to know what the house was made of, what specific appliances and furnishings were in the kitchen and how it was laid out, such that the heat spread exactly as it did in this instance. And she would want to know this in terms general enough to apply to all instances of the spread of flames. The pyrologist is looking for the materials to inform her scientific theorizing about fire, and the tools for such theorizing are not the same as the tools that allow us to achieve a common-sense grasp of the way fires work. The homeowner or neighbor may be interested in preventing future fires, and so he or she would be interested in making sure that there are no gas leaks in the neighborhood or that the house was not made of shoddy materials. But the pyrologist is interested in understanding and preventing fires much more generally, and the level of detail at which her background understanding frames the relevant "Why" question is much more fine-grained than the background understanding that frames the "Why" question posed by the homeowner or neighbor.

The conclusion to draw from these observations about our explanatory practices in relation to the first principle assumed by van Lommel and Alexander, *Complete Explanation*, is that these authors are concerned with a specialized form of explanation. They want to explain near-death experiences in such a manner that they can construct maximally general theories about why they occur, theories that cover all aspects of all cases. Given this explanatory project, this principle seems like a reasonable requirement.

While there may be good reason for van Lommel and Alexander to accept *Complete Explanation* as a requirement on adequate explanations of near-death experiences, we think that the other principle they implicitly accept, *Single Explanation*, is manifestly without merit. It should be obvious that *Single Explanation* is not a requirement for everyday explanations. More than one factor may be invoked to plug the gaps in our understanding of why an event, such as a fire, occurred. We appeal to the stovetop flame or the lightning strike as circumstances call for. The same goes for everyday explanations of near-death experiences. But as we have taken van Lommel and Alexander to be engaged in a more specialized explanatory project, we need to show that, upon reflection, it is equally clear that *Single Explanation* is out of place even in the context of the kind of specialized explanations that concern them. We need to argue that even scientific explanations should be allowed to invoke multiple explanatory factors. That shouldn't be too difficult.

When you think about the phenomena we are interested in explaining, the problem with *Single Explanation* seems obvious. Near-death experiences are complicated, encompassing various general kinds of experience (out-of-body experiences, life reviews, and so forth) that manifest in diverse forms in diverse individuals (people in cardiac arrest, under general anesthesia, and so forth). To put it another way, not all near-death experiences and not all near-death experiencers are alike in every respect. Given the project of providing a maximally general theory of such diverse phenomena, it would seem highly counterintuitive to insist that adequate explanations of them cannot differ in certain respects. But this is precisely what one does if one accepts *Single Explanation*.

The trouble with *Single Explanation* in this context may be brought out further by analogy. Suppose you went to your doctor because you had a fever, runny nose, and earache. After checking you out, your doctor says that she cannot explain why you are experiencing these symptoms because there is no single factor that she can find that explains them all. Perhaps the earache can be explained by this one kind of bacterium she found in a culture from the swab she took. But the presence of these bacteria, she says, would not explain your runny nose. Moreover, she has seen various patients recently with similar symptoms, and she cannot find any one thing that can explain the suite of symptoms in all of you.

It would be absurd for your doctor to turn a blind eye to the possibility that your several symptoms have several distinct causes. Similarly, it seems obvious that she should not suppose that the same suite of symptoms in different people cannot have different underlying causes, such as a cold virus and a bacterial infection. It seems obvious that the doctor in this hypothetical scenario is invoking an absurd standard for explaining her patients' symptoms. But she is simply invoking *Single Explanation*.

Reflection on this scenario (or any number of similar scenarios) reveals the inappropriateness of this principle even in specialized contexts. Surely neither van Lommel nor Alexander, both of whom are practicing MDs, would invoke *Single Explanation* in the context of their medical practices. So why would they invoke it in the context of explaining near-death experiences? If the principle is unfit in one context, why invoke it in the other? *Single Explanation* ought to go.

If we reject *Single Explanation*, what should take its place? As a first pass, we should allow for disjunctive explanations. An adequate explanation might take the following form: near-death experiences are caused by x or y or z, where x, y, and z stand for diverse factors that may explain various features of near-death experiences. An example (and, we stress, this is just an illustrative example, not a proposed explanation of near-death experiences) might be as follows: near-death experiences are caused by lack of oxygen or high carbon dioxide or oxygen deficiency or the use of certain drugs. Rather than considering each of these factors in isolation, as van Lommel does, we might consider them as a set. The same goes for the set of nine factors considered by Alexander.

By allowing for disjunctive explanations, we allow, first, that different aspects of near-death experiences might have different causes and, second, that the same aspects might have different causes in different instances. Near-death experiences do not all arise in the same ways or under the same conditions. And they do not all include the very same features, even if there are some general characteristics that many do share in common. Given that not all near-death experiences are identical, why think that they are all created in identical ways?

In addition to disjunctive explanations, we should allow for the possibility that various factors may explain near-death experiences in combination with each other. We should allow for explanations that appeal to conjunctions of explanatory factors, even conjunctions that work synergistically. These kinds of explanations are nothing new. Consider again the case of the house fire. The burner on the stove did not cause the fire in isolation. It required flammable materials in a particular arrangement, a breeze or draft and the presence of oxygen to set the house ablaze. It would be absurd to insist, just because a lit gas burner on a stove could not, by itself, cause an entire house to catch on fire, that this burner in combination with several features of its environment could not have caused the house fire. Clearly, the presence of additional factors worked in combination with the lit burner to engulf the house in flames. And these additional factors are not simply to be subsumed under the rubric of the background factors assumed to be a part of the general understanding that makes the relevant "Why" question intelligible. The specific arrangement of the appliances and furniture in a given kitchen, or the presence of a draft on a given day, are not a part of the background understanding that makes it possible to ask why the house burned down. They are factors that work together to explain why the house burned down as it did.

To drive home the point, consider your reaction if the investigators looking into a fire at your house were to tell you that, because a gas burner, by itself, does not cause a house fire, they were certain that this gas burner left turned on and unattended on your stove could not have caused the fire in your house. Surely you would find their reasoning absurd. Perhaps you would even ask to speak to the chief, if only to request that a different team be sent out to investigate. Our point is this: the fact that a single factor, in isolation, cannot explain a given phenomenon does not support the claim that this factor cannot serve as a part of the explanation, in combination with other factors.

Surprisingly, van Lommel and Alexander seem to be guilty of just this kind of faulty reasoning. They consider several physical factors in isolation, concluding of each one that it cannot do the requisite explanatory work all on its own. Then they conclude that each factor is explanatorily irrelevant with respect to near-death experiences. This argumentative strategy ignores the possibility of conjunctive, or even synergistic, physical explanations of near-death experiences. For example, Alexander argues that DMT (a naturally occurring hormone that can cause hallucinations) does not, all by itself, adequately explain near-death experiences. But this does not establish that it is unable to do so in combination with other factors, such as, for instance, distorted recall of memories and high levels of activity in certain neuronal networks. Let's be clear. We are not here claiming that this particular combination of factors can explain near-death experiences. Establishing that would require scientific expertise we do not possess. We are making a point about what constitutes an adequate explanation; the possibility that this combination of factors adequately explains near-death experiences is not ruled out by showing that any one of these factors is explanatorily inadequate in isolation from the others. Furthermore, we claim that, due to his allegiance to *Single Explanation*, Alexander is blind to this possibility.

This is true even if it turns out that Alexander's use of this principle is merely implicit and not conscious. Reasoning can be faulty, even if one does not have the principles behind one's thinking fully in view. Indeed, this sort of situation would appear to *enhance* the prospect of faulty reasoning. It is easier to be guided by faulty principles when one is not reflectively aware of the principles guiding one's thought. So we are not here claiming that either Alexander or van Lommel explicitly endorses *Single Explanation*. Rather, examination of their reasoning reveals that their thinking is, in fact, even if only implicitly, guided by this principle. And furthermore, this is a bad principle. It should play no role in directing one's pursuit of explanations in general, let alone the sort of specialized explanations of near-death experiences at issue here.

To recap, we have been arguing in favor of multi-factor explanations of different kinds in various explanatory contexts, including near-death experiences. We think that looking into the prospects of explaining near-death experiences by appeal to multiple physical factors is very worthwhile. We are not the only ones who think so.

Kevin Nelson is a practicing neurologist with a keen interest in spiritual experiences, especially near-death experiences. And he has been collecting data for some time now in support of an explanation of near-death experiences that appeals to multiple physical factors.[5] He argues that we can explain the various features commonly reported by subjects of near-death experiences by appealing to factors like changes in blood flow during crisis and the blending of different states of consciousness, such as that associated with waking consciousness and that associated with the rapid eye movement

---

5. For a very nice and succinct overview of some of his work related to near-death experiences, see Nelson (2014). Sacks (2012a: 260–261) discusses some of Nelson's work as well.

(REM) sleep cycle. Importantly, Nelson not only points to the possibility that the physiological mechanisms he cites are potentially explanatory of the phenomena associated with near-death experiences, but he also seeks to explain why these mechanisms would be triggered in near-death episodes—for instance, because they are connected to our fight or flight response—and also why their manifestation in these contexts would be evolutionarily advantageous. And he connects the dots between the ways in which our brains function and the ways in which the brains of our close ancestors on the phylogenetic tree function. This is precisely the sort of research program we think promising.

In addition to the kind of multi-faceted approach to the explanation of near-death experiences in physical terms exemplified in Nelson's work, various social and contextual factors are relevant too. Recall that the out-of-body experiences of Pam Reynolds and the man with the dentures might well be influenced by various factors that were not, strictly speaking, part of the original experience. We cited, for instance, the possibility that the man had earlier seen the nurse he later identified as the one who removed his dentures during his stay in the hospital. The experience he then described included the visual experience of seeing this nurse, and yet this aspect of the experience may have been a late addition. This is one kind of environmental factor that might be relevant to explaining particular near-death experiences: the influence of later events on recollected earlier events.

A second set of factors might be thought to "frame" and "structure" the report of a near-death experience. Consider that in pretty much all Western cultures we have the saying, "There's always light at the end of the tunnel." This idea gives us hope even in the direst circumstances. It is possible that this inspirational metaphor might frame and structure one's report of a prior experience of light against a dark background. It is possible, that is, that one might report seeing a light in a dark tunnel, where this report is based on a mixture of raw experience and background information, including this widespread metaphor. We have contended that the proper explanation of even just one aspect of a near-death experience may invoke various factors in combination. And it would not be surprising if some of these factors were culturally specific framing conditions on how we interpret our experiences. Thus, we might find that complete explanations of near-death experiences appeal to multiple factors, some of which are physiological and others social and contextual.

This point is supported by apparent cross-cultural differences in the contents of reported near-death experiences. In Japan some people who have had near-death experiences report seeing themselves in a rock garden. It turns out that in Japan, rather than the metaphor of light at the end of the tunnel, there is a popular idea of buying and tending a rock garden with one's family and friends in old age.[6] This is a different story that gives hope at the end of life. We see no reason to think that the human mind could not have mechanisms that allow comforting ideas to frame and structure our interpretations of what we experience in fraught moments, such as near-death episodes.

6. We thank Bruce Greyson for mentioning these fascinating points in a presentation to the International Board of Advisors of the John Templeton Foundation, June 2010, and in personal correspondence.

Chapter 9

Are Simpler Explanations
More Likely to Be True?

The various aspects of near-death experiences form a motley crew. They include out-of-body experiences, life reviews, communication with the deceased, tunnels of light, special states of consciousness, and more. The variety of things to be explained is part of what motivates our proposal that multi-factor explanations be taken seriously. Why reach for a one-size-fits-all explanation of such multifarious phenomena? The single-factor strategy just doesn't seem appropriate.

But not everyone agrees. E. W. Kelly, Bruce Greyson, and Ian Stevenson have offered an interesting argument for the claim that the most plausible explanation of near-death experiences will be nonphysical. They identify three factors—enhanced mentation (or superior cognitive activity), perceiving one's physical body from another location, and extrasensory perception—that they admit do not individually provide significant reason to conclude from near-death experiences that "consciousness can function independently of the physical brain and body." But, they argue, when we find these factors all together, this may be taken to provide evidence that physicalism is false:

> When taken singly, each of the three features we have mentioned is open to an explanation other than the survival of consciousness after death. First, episodes of enhanced mentation occur in other conditions, such as during periods of creativity, manic episodes, or while a person is under the influence of mescaline and some other drugs. It is surprising, however, to hear reports of enhanced—or even of normal—mentation from persons who were ostensibly unconscious and often close to death; one would expect the cognitive functioning of such persons to be absent, seriously impaired, or at best diminishing. Nevertheless, some common factor may eventually be shown to stimulate the cognitive processes of persons in a wide variety of physiological conditions, although no such factor adequately accounting for all near-death experiences has yet been identified. As for the experience of looking down on one's body from above, this might be explained as a hallucination generated by psychological or somatic conditions associated with illness or injury [references suppressed]. Finally, experiences of extrasensory perception, like enhanced cognition, are reported to have occurred to many persons who were in good health and far from being near death. Although no adequate explanation—normal or paranormal—has yet been proposed for such experiences, the explanation may not require a component of human personality capable of surviving bodily death.

When, however, we consider the occurrence of the three features *together*, and have to begin invoking different explanations for the different features, we should ask whether a single hypothesis—that consciousness can function indepen-dently of the physical brain and body—might be better able to explain all these features.[1]

The basic idea here seems to be that, even if we can explain the various elements of near-death experiences, in isolation, by appeal to various physical factors, it might still make for a better expla-nation of these near-death experiences, overall, if we appeal to a single nonphysical factor. This argument poses a challenge to the strategy we argued for in Chapter 8, as it calls into question the explanatory adequacy of multi-factor physical explanations. But we think this argument should be resisted for at least two reasons.

The first is that it appears to rely on yet another objection-able principle. The move in the above passage from considering multiple physical factors as explanations of the various features present in these near-death experiences, in the first paragraph, to considering just one explanatory factor, in the second paragraph, depends on favoring a single-factor explanation to a multi-factor explanation. In the absence of this preference, there would seem to be no reason to pose the question these authors think we should. Thus, this argument appears to rely on a revised ver-sion of *Single Explanation*. We might put the revised principle as follows:

Simple Explanation: Given two rival explanations of near-death experiences, we should prefer the one that invokes *fewer* explanatory factors to the one that invokes *more* explanatory factors.

1. Kelly et al. (1999–2000: 515). See also Cook et al. (1998).

We argued, in Chapter 8, that *Single Explanation* ought to be rejected. We think that this revised version, *Simple Explanation,* should be shown the door as well.

Our main objection to *Simple Explanation* is that it ignores the relevance of truth. Presumably, we want our explanations to be true, and not just elegant. But this principle does not count truth as relevant to which explanations are to be preferred. It simply instructs us to prefer simpler explanations. And there is no reason for thinking that simpler explanations are, as such, more likely to be true than more complex explanations. Indeed, the opposite often seems more likely. As we noted in Chapter 8, it seems to be generally true that various physiological conditions familiar from experience, such as the suite of symptoms commonly associated with the flu, are capable of being explained by multiple factors rather than one factor on its own. Why suppose that near-death experiences are any different in this regard?

Of course, there are *pragmatic* reasons to prefer simpler explanations, all other things equal. But these are not reasons to suppose that simpler explanations are more likely to be *true.* The well-known philosopher of science, Bas van Fraassen, makes this point nicely:

> Simplicity is quite an instructive case. It is obviously a criterion in theory choice, or at least a term in theory appraisal. For that reason, some writings on the subject of induction suggest that simple theories are more likely to be true. But it is surely absurd to think that the world is more likely to be simple than complicated (unless one has certain metaphysical or theological views not usually accepted as legitimate factors in scientific inference). The point is that the virtue, or patchwork of virtues, indicated by the term is a factor in theory appraisal, but does

not indicate *special* features that make a theory more likely to be true (or empirically adequate).[2]

There are various reasons for preferring simple to complex theories, other things equal. To take just one, simple theories are easier to comprehend. And it is difficult to test or base predictions on a theory one does not understand. But there are also compelling reasons to think that at least some aspects of reality are complex. Adequate explanations of these aspects of reality should capture this complexity. And there does not seem to be a compelling reason to think that such explanations will not themselves be complex. In other words, there is no compelling reason to think that the world is likely to aid in our quest to comprehend it by always allowing itself to be captured by simple explanations. And more to the point, given the apparent complexity of the various phenomena associated with near-death experiences, there does not seem to be good reason to suppose that we are looking for a simple explanation of them.

The point we are making here is similar to our earlier point about sincerity. Sincere testimony grabs our attention. We are more likely to take seriously the evidence an apparently sincere witness purports to give us. But even a sincere defendant needs a good alibi. Sincerity is not itself evidence of the truth of what one is saying. Granted, simplicity does seem to be an explanatory virtue. We seek explanations because we want to comprehend the way things work. And the simpler the explanation, the easier it is to achieve a mental grasp of what is going on. But we do not just seek a grasp of the way things work; we seek *true* accounts of the way things work. And it may be that the true account is not all that

2. Van Fraassen (1980: 90).

easy to understand. It may take real work to grasp the wonders of the world. Just as the sincerity of a piece of testimony is not itself evidence of the truth of a given report, simplicity is not itself evidence of the truth of a given explanation.[3]

Again, we are not denying that there is something to be said for simpler explanations. They are easier to understand, and so they better satisfy our aims in seeking and providing explanations. But there is a danger in appealing to a principle such as *Simple Explanation* that instructs us to prefer simpler explanations to more complex ones: a focus on simplicity may get in the way of arriving at the truth. We may be prone to underestimating the merits of certain explanations just because they are complex, and we may be prone to overemphasizing the merits of other explanations just because they are simple. These tendencies are especially insidious when coupled with other well-known tendencies of ours, such as those of discounting evidence that tells against our current beliefs and seeking evidence that tells in their favor. We will explore the role of confirmation bias in Chapter 12. But it is worth remarking now on the facilitating role a principle such as *Simple Explanation* might play in helping us to shore up certain of our cherished beliefs.

Consider someone who believes in an omnipotent, omniscient, and omnipresent God. She may be fully comfortable, and indeed inclined, to invoke God, or some aspect of God's nature, to explain why she had the experience of seeing deceased relatives in heaven during her time in a coma. She saw these people in heaven because, through the grace of God, they traveled there after their deaths and God wanted her to know that she is still loved by them. In a way, this is a rather simple explanation of

3. For more in this vein, see Ball (2014).

this woman's near-death experience. It invokes the work of one entity, God. And it certainly seems simpler than an explanation invoking various factors, such as a fear of death, the brain's capacity to store information and construct experiences from it after the fact, and so on. In short, this explanation in terms of God seems to be a simpler one than the kind of multi-factor, physicalist explanations we think should be taken seriously. Moreover, it is an explanation that serves to confirm this woman's religious convictions. If God really did take her to heaven, then God must be real and heaven must be a place where people like her go when they die.

This supernatural explanation of her near-death experience may appeal to this woman both because it coheres with and supports her deeply held religious beliefs and because it is simpler than rival, physicalist explanations. But neither of these aspects of the explanation is evidence that it is *true*. We have already provided reasons to think that true explanations are not always simpler than their rivals. And though this is certainly contentious, it bears pointing out that no matter how deeply and sincerely held the woman's religious beliefs may be, it is possible that they are mistaken. (Think about it: there are many different and incompatible religious traditions, so even if one of them is the true view, not all sincerely held religious beliefs can be true.) The fact that a given explanation coheres with her religious beliefs clearly does not, in its own right, count in favor of the truth of that explanation. If this woman wants a true and accurate understanding of what she experienced, then she should be open to the possibility that her experience tells against some of what she thought she knew about the world around her. Indeed, this is of a piece with the supernaturalist's counsel to the physicalist: accept that near-death experiences blow up your worldview!

We have been responding to an argument by Kelly, Greyson, and Stevenson to the effect that we should take supernatural explanations of near-death experiences seriously on the grounds that they provide simpler, and thus better, explanations of the relevant phenomena. Our first response is that the principle, *Simple Explanation*, on which this argument rests, should be rejected. Simplicity is not a reliable guide to truth. But we have a second response to their argument as well. Put simply, the challenge is this: why think that supernatural explanations are simpler than their physicalist rivals? The kind of explanation that Kelly, Greyson, and Stevenson would like us to take seriously involves the interaction between our physical bodies and a nonlocal consciousness capable of surviving the death of one's body. How, exactly, is this supposed to provide a simpler explanation than one involving multiple physical factors?

A supernaturalist who accepts the existence of both the physical and the nonphysical faces the daunting task of explaining how the nonphysical is supposed to interact with the physical. It is mysterious how nonphysical factors or phenomena could interact with our physical bodies and brains at all. For instance, how is an immaterial consciousness supposed to receive information from the eyes in order to give rise to visual perceptions? It is even more mysterious how the nonphysical could interact with the physical in a manner that would allow us to appeal to nonphysical factors in order to explain the particular phenomena associated with near-death experiences. How is the immaterial consciousness supposed to exit and then reenter the body before and after an out-of-body experience? In order to arrive at an adequate and useful explanation that appeals to nonphysical factors, we have to fill in the details. Thus, even if we do take supernatural explanations of near-death experiences seriously, it is not at all clear that we are thereby simplifying things. Even though adequate multi-factor, physical explanations may need to be quite complicated, there are grounds for thinking that any acceptable supernatural explanation would need to be even more complicated.

Consider what it would take to provide an adequate and complete supernatural explanation of near-death experiences—an explanation that situates them within the broader context of our ordinary lives. First, we would have to explain how the relevant nonphysical factors explain the phenomena of near-death experiences. Second, we would need to understand how these nonphysical factors *interact with* our physical bodies in *normal* circumstances. If we are appealing to an immaterial consciousness, then we need to understand how this could give rise to a conscious experience of seeing one's own body during cardiac arrest and also how it could give rise to normal, everyday visual experiences involving our eyeballs. The problem here is a special case of explaining how the nonphysical and physical realms can have causal interaction. Finally, an adequate and complete supernatural explanation would have to make sense of the fact that people recall and describe their near-death experiences later, when their physical bodies are in good working order. By all accounts, these recollections and descriptions involve the brain, mouth, and so on. But how, then, are these physical processes supposed to relate to the nonphysical mechanisms appealed to in the explanation of the original experience itself? Satisfying these three criteria is a tall order.

Once we realize that there are further pieces of the puzzle to put in place in order to flesh out a proposed single-factor supernatural explanation, it seems doubtful that it will be any simpler than a rival multi-factor physical explanation. So, if simpler explanations are what we are after, perhaps we ought to stick to the physicalist paradigm. Multi-factor physical explanations wear their complexity on their sleeves, as it were. But they have the virtue of cohering with a vast body of scientific knowledge. It is easier to fit them into our broad, common-sense understanding of the way the world works. Supernatural explanations appear simpler, but this appearance vanishes once we try to fit them into the bigger picture. In trying to understand how the various pieces of the puzzle are supposed to fit together, we end up finding more gaps in our understanding that need filling in.

We can make the point more concrete by considering an example. Pim van Lommel appeals to the notion of a nonphysical consciousness received and transmitted by one's physical brain in order to explain the possibility of recording visual perceptions from a position outside one's body and reporting them later, after one's consciousness has returned to the confines of one's body.[4] Let us suppose there is no special problem understanding how a nonphysical consciousness can receive visual perceptions outside the spatial location of a physical body. Even so, the point we have been pressing is that we still need an explanation of how this nonphysical consciousness interacts with the body in order for the brain to receive and transmit its signals. Invoking this nonphysical factor may appear to allow one to explain the abnormal elements of the near-death experience—such as disembodied, conscious visual perception—but at the cost of creating the problem of explaining normal elements of everyday experience, such as later recalling the visual experiences one has had and reporting them during moments of normal consciousness. Again, this is an instance of the general problem of explaining the possibility and

4. Compare van Lommel (2010, 2013).

nature of causal interaction between the nonphysical and physical realms—a notoriously daunting problem. If simplicity is what we are after, then it is not at all clear that van Lommel's theory of a nonlocalized consciousness fits the bill.

The complication with supernatural explanations has to do with the need to explain the interaction between the physical mechanisms in terms of which we understand most of our everyday experiences and the nonphysical mechanisms said to be part of the explanation of near-death experiences. This appears to be a specific instance of a general problem for those who deny physicalism about the relationship between the mind and the brain. Like other dualisms, it is a part of the doctrine of supernaturalism to hold that there are two kinds of things: physical things and nonphysical things. And this poses the problem of how to explain causal interaction (or *any* interaction) between these two kinds of things. It is unclear, for example, how the brain and central nervous system can interact with the mind, conceived of as nonphysical. How does the mind induce movement in the body? How does your intention to lift your coffee mug to your lips, conceived of as an element of your nonphysical mind, result in the movement of your arm, which is a physical part of your body caused to move by physical processes involving neurons, muscles, and so on? The issues here are well worn, and the debate is far from settled.[5] Thus, it is not only unclear that supernaturalism provides simpler explanations, but it is more generally unclear whether it presents a satisfactory explanatory framework at all.

We have good reason to avoid appealing to the nonphysical in explaining near-death experiences. Though admittedly

5. For a nice overview of the main issues, see Robinson (2012).

incomplete, our understanding of the physical world is quite good. And we have fruitful methods for improving upon our understanding of such matters. Given the problems facing the supernaturalist, it seems wise to stick with the task of coming to a complete understanding of things in wholly physical terms. This includes the special context of explaining near-death experiences. Better, we claim, to search for ways to combine multiple physical factors into adequate explanations of the phenomena than to open up a new can of worms.

Not everyone, however, will find this line of reasoning persuasive. As noted by van Lommel, Alexander, and others, all single-factor physical explanations seem inadequate as comprehensive explanations of near-death experiences. But some might insist that scientists are quite familiar with the various possible single-factor physical explanations of near-death experiences, *as well as with the ways in which these single physical factors could combine.* And so we might conclude that it is unlikely that *any* physical explanation of near-death experiences, even a multi-factor one, would be adequate.[6] This is, in effect, to flip the above line of reasoning on its head. One might think that the thing to do is work to overcome the age-old difficulties facing the denial of physicalism, as opposed to putting our stock into the prospects of the physical sciences. Perhaps supernaturalism should be taken seriously because we have reason to believe that the prospects of physicalism for explaining near-death experiences are, ultimately, hopeless.

The problem with this response is that it seems hasty. The relevant sciences are works in progress. For example, neuroscience is in its infancy, and there is much that we do not know about the

6. We would like to thank David Chalmers for raising this objection in personal correspondence.

neurophysiological processes in the brain and how they under-write mental phenomena and experiences. It is not implausible that we have not yet identified certain candidates for physical factors that could help to explain consciousness more generally, and near-death experiences in particular. And it seems premature to suppose that we are at a point in the development of neuroscience (or any other relevant field of scientific inquiry) at which we could be confident that we have a good grasp of how known physical factors might interact. We should reserve judgment on the prospects of physicalism given the evident and expected progress of science.

Moreover, given the daunting problems just mentioned of explaining how the physical can interact with the nonphysical, it would seem judicious to suppose that it is more likely that we will develop an adequate physical explanation of near-death experiences than that we will develop an adequate nonphysical explanation of near-death experiences. Even if physicalism faces problems—for example, the problem of explaining how, precisely, the brain is supposed to produce consciousness—solving these problems seems more tractable than solving the problems facing the supernaturalist. Making sense of supernaturalism requires positing two fundamentally different kinds of things—the physical and the nonphysical—and working out how to characterize them individually and how to make sense of their interaction. Making do with physicalism, by contrast, requires putting just one house—the physical—in order. This task appears less daunting.

Near-death experiences are perplexing phenomena that, so far at least, defy our quest for complete understanding. In examining the issue of whether any adequate explanation of the phenomena must appeal to the nonphysical, or whether there may be adequate physical explanations of near-death experiences, our contention is that we would do well to consider the *overall explanatory progress* that would accompany either strategy. We may not, at present, be able to see our way to a physical explanation of near-death experiences, but that is no good reason for pronouncing dead the hope of explaining them in wholly physical terms. Indeed, given that the relevant sciences are in their infancy and that the physical sciences provide us with an ever-expanding body of knowledge, it would seem reasonable to assume that our current state of knowledge is not representative of physicalism's prospects. Together with a recognition of the daunting general problems facing the denial of physicalism, and the specific problems facing those who would invoke nonphysical factors in order to explain near-death experiences, we take these considerations to tell in favor of physicalism. Especially given the general problems of explaining how the nonphysical and physical realms can interact, we contend that physicalism is overall the *better* theory than supernaturalism. To return to our judicial metaphor: the evidence presented to the jury does *not* provide a compelling case for supernaturalism, even when the standard of proof is less stringent than *certainty* or *near-certainty*. Supernaturalism is not even *probably* true.

At this point you may be nodding your head in agreement, while at the same time rolling your eyes in disappointment. We have, you might think, missed the whole point of the argument by Kelly, Greyson, and Stevenson with which this chapter began. You might think that we have saddled them with a claim they do not make. Nowhere do they explicitly invoke a *nonphysical* explanatory factor. Rather, they suggest that the confluence of the three factors they cite provides support for investigating the possibility of a consciousness that survives after one dies. And there is nothing incoherent about claiming that such a consciousness is physical. Thus, even if one obvious interpretation of their proposal is that they are appealing to a *nonphysical* factor—a nonlocal, nonphysical consciousness—in order to explain near-death experiences, they may be appealing to a *physical* factor that is not an element of our current ordinary understanding of the physical world and how it works. They might be appealing to a physical factor that is not localized in the body and that we have not yet discovered through empirical investigation.

But this does not really pose any challenge to the argument of this chapter. For even if the single factor in question is a *physical*, nonlocal consciousness, it raises many of the same questions we have been discussing. The basic idea behind the worry we have been pressing against the denial of physicalism is that appealing to a nonphysical factor raises a host of issues about how this factor is supposed to interact with familiar physical factors. This worry would apply as well to a proposal that cited a mysterious physical factor. Appealing to a nonlocalized, physical consciousness in order to explain these aspects of near-death experiences does not make any clear explanatory progress for the same reasons that appealing to a nonphysical consciousness did not.

Given a nonlocalized, physical consciousness, we should want to know how our ordinary conscious experiences are to be understood. If our consciousness can function independently of our physical brains and bodies then we should want to know how it can function *in tandem* with our physical brains and bodies as well. How does this nonlocal consciousness interact causally with our brains, located, as they are, in our skulls, so as to produce the observable effects that scientific investigation has revealed? And why have we been hitherto unable to observe this nonlocal consciousness or pick it out as distinct from other observable phenomena? Moreover, what kind of physical thing is this consciousness supposed to be, such that it can function independently of the matter making up our brains and is not itself observable, by current means, in the environment around us?

The same points we raised in response to the proposal that we appeal to a single, nonphysical factor seem to apply equally well to this proposal involving a nonlocal, physical consciousness. First, an explanation in terms of this factor is not obviously simpler than an explanation in terms of multiple physical factors. Second, it is not at all clear that an explanation in terms of this mysterious physical factor makes any explanatory progress.

Once again, we can illustrate these points by considering van Lommel's account of consciousness, which he describes in terms of an analogy with radio waves. He suggests that we might think of our brain as like a radio transmitter and consciousness as like radio waves. Earlier, we considered this proposal as a version of supernaturalism, because that is the most natural interpretation, given other things van Lommel says. But the analogy lends itself to a physicalist interpretation as well. Radio waves are a well-understood physical phenomenon, and we understand how our radios receive these waves and then produce sounds via speaker systems. That is, we understand *the physical mechanism* whereby our radios receive the radio waves and also transform them into sounds intelligible to listeners. But our understanding of how radios and radio waves interact does not help us to make sense of how the brain and consciousness, understood by analogy to a radio and radio waves, are supposed to interact. In other words, the analogy is simply not illuminating. It may be suggestive, but it does not shed any light on how consciousness works. The trouble is that we do not know how to make sense of the interaction between the mysterious physical factor, nonlocalized consciousness, and the more familiar gray matter in our skulls. What is *the mechanism* in virtue of which our brains are supposed to receive the nonlocalized "consciousness waves"? Absent even a preliminary grasp of this mechanism, it remains completely mysterious how understanding consciousness as nonlocalized and physical is supposed to shed any light on the nature of our experiences, including near-death experiences.

No matter whether we take it to be positing a physical or a nonphysical single-factor explanation of near-death experiences, the explanation proposed by Kelly, Greyson, and Stevenson raises more questions than it answers. By contrast, the multi-factor explanation they think it should replace is nested in a (more or less complete) general understanding of how the physical elements making up the world hang together. For this reason it does not raise these same questions. Thus, we conclude, their single-factor explanation is not really simpler than the multi-factor one. It only appears simpler if you ignore the thorny, complex issues it raises. Even if you prefer simpler explanations, there is good reason to reject both supernaturalism and explanations that involve mysterious physical factors.

Chapter 10

Near-Death Experiences, Transformation, and the Afterlife

We have been focusing on issues of explanation. Why do people have near-death experiences? What is the role of our brains in these experiences? But questions about their causes do not exhaust the topic's interest. Many have noted the profound and transformational nature of near-death experiences—especially, the people who have had them and shared their stories! We want to understand this too. How are we to make sense of near-death experiences in light of their significance for those who have them?

The data on the transformative effects of near-death experiences are not just anecdotal. In a well-known longitudinal study, van Lommel led a team that interviewed several hundred patients at Dutch hospitals who had been successfully resuscitated. They talked to people who had survived cardiac arrest. They found that 18% of these people had had near-death experiences, or experiences that were similar in various ways, and they were able to gather information about the effects of these experiences on many of their lives. They interviewed 74 of these patients two years, and 23 of them again eight years after their initial resuscitations. They found that, in relation to a control group of subjects, patients who had had near-death experiences were significantly transformed. After two years, "people who had [an] NDE had a significant increase in belief in an afterlife and decrease in fear of death compared with people who had not had this experience."[1] And after eight years,

> they had become more emotionally vulnerable and empathic, and often there was evidence of increased intuitive feelings. Most of this group [of NDErs] did not show any fear of death and strongly believed in an afterlife. Positive changes were more apparent at 8 years than at 2 years of follow-up.[2]

Having a near-death experience was correlated with higher spirituality and pro-sociality. Those with near-death experiences were not only more likely to believe in an afterlife, but they were also more likely to exhibit acceptance and love of others. It is hard to argue with the characterization of (at least some of) these effects as "positive changes." Subjects of near-death experiences were transformed for the better. And this was not just due to their having survived cardiac arrest. Their transformation was measured in contrast to other survivors, who had apparently not had near-death experiences.

There are various ways of interpreting the significance of these data. Putting these findings about the transformative power of near-death experiences together with the lack of satisfactory physical explanations of the phenomena, van Lommel and his colleagues call for serious consideration of alternatives to the view that consciousness is localized in the brain. But others have taken

1. Van Lommel et al. (2001: 2042).
2. Van Lommel et al. (2001: 2043).

the transformative power of near-death experiences to suggest something more. For example, Long writes:

> The fact that near-death experiences bring about transformation is powerful evidence of the afterlife. For me it's evidence that those who step briefly into the afterlife bring back a piece of it when they return.[3]

To be frank, this is a leap of logic and simply unwarranted. Why exactly would the transformational character of near-death experiences imply that they put people in contact with a nonphysical realm?

The first thing to note is that just as we need to distinguish between the reality and the accuracy of an experience, including a near-death experience, we also need to distinguish between the profoundly transformative nature of an experience and its accuracy. Consider the phenomenon of hallucinations due to Charles Bonnet syndrome. Oliver Sacks details the story of one woman, Virginia Hamilton Adair, who experienced her hallucinations as profoundly inspiring.[4] They became her muse. And she openly referred to them in several acclaimed collections of poetry published late in her life. Adair experienced her hallucinations, many of which she had after going completely blind, as powerful and transformative. It is not a stretch to say that their effects on her were, in certain ways, similar to the effects of near-death experiences on some of those who have them. But Adair's hallucinations were due to a syndrome that is manifestly explicable in terms of the physical sciences. And though no one, presumably, would take

---

3. Long (2010: 197).
4. Sacks (2012a: 32–33).

Adair's experiences to correspond to an independent reality, they were deeply moving and inspiring. Thus, the transformative and inspirational power of an experience in itself simply does not warrant conclusions about its accuracy. Nor does it rule out the appropriateness of explaining that experience in physical terms.

As with our analysis of the similarity of near-death experiences the world over in Chapter 7, we need not here look to a link with a nonphysical realm to get a plausible explanation of the transformative character of these experiences. First, we can look to their context. Near-death contexts are, by definition, situations of mortal danger. These are unique moments in a person's life. When reflecting on one of these experiences after the fact, one could be expected to reflect on the context surrounding it and to attach great significance to the fact that it was a *near-death context*. It would be natural to attach significant weight to a near-death experience simply in light of when it occurred. The profoundness of the moment may naturally affect the profoundness of the experience. Further, the *contents* of near-death experiences themselves are often deeply meaningful. Witnessing one's own resuscitation or experiencing a reunion with deceased relatives is a profound matter in itself. As such, it would be natural for it to resonate deeply and to leave a lasting impression, including changes in one's orientation toward life.

These considerations may help us to make sense of the power of near-death experiences to transform the individual without assuming that they are *accurate*. It may be that the transformation is propelled by the content of the subject's experience as it presents itself in a near-death context. And whether the content of this experience matches an external reality may be irrelevant to its effects on the individual. A physicalist-friendly approach here is *at least as plausible* as one that claims that the individual has somehow returned after having had direct contact with a nonphysical realm. Indeed, a stronger statement is warranted here. Given the problems noted in Chapter 9 about the interaction between the nonphysical and physical, it significantly understates the case in favor of physicalism to claim that it is on a par with supernaturalism.

Various people have pointed out that part of the transformation that typically occurs after a near-death experience involves a heightened sensitivity to the needs and interests of others—a kind of heightened moral sensibility. Van Lommel summarizes this dimension of the transformational nature of near-death experiences as follows:

> After the NDE, relationships with others change noticeably, and people are now capable of greater compassion.
>
> People are more forgiving, more tolerant, and less critical of others. . . . Appreciation of relationships increases; people spend more time with family, friends, and relatives, and they are more willing and able to share emotions with others. They are more compassionate and caring and set greater store by unconditional love. . . . A greater sense of justice is coupled with the urge to tell the truth and say what is on one's mind. Any trace of past aggression is usually gone. It is replaced by the need to help and support others, which usually leads to a career change in favor of the care professions, such as nursing, care of terminal patients, or voluntary work with elderly people or low-income families. NDErs are also more likely to donate to charities or to dedicate themselves to a social cause.[5]

5. Van Lommel (2010: 53).

Of course, increased compassion for others—increased moral sensibility—is admirable and noteworthy in anyone, including those who have had near-death experiences. This is certainly a most welcome feature of the transformations that typically accompany them. But it is worth reflecting on the relationship between a near-death experience and this sort of heightened moral sensibility or orientation toward others.

Whereas increased moral sensibility certainly *can* and often *does* happen as a result of a near-death experience, note that it in no way *requires* a near-death experience. Nor does it require that such an experience be interpreted as involving nonphysical mechanisms or contact with a supernatural realm. People become more compassionate and caring for various reasons. They learn from their own mistakes; they hit rock bottom; they follow sage advice. Sometimes people just see the light, so to speak, and make changes for the better for no discernible reason. Our characters are not set in stone. Some of us can, and do, become better people—even some of us who have never had a near-death experience. Moreover, whether or not we *do* become better people, we certainly *should*. Not having had a near-death experience is not an excuse for being a bad person. Neither is not believing in heaven. Being attentive to the needs of others is presumably required of us as members of the moral community. Even those of us who have never had a near-death experience should work to improve our moral sensibilities. These points are worth keeping in mind, especially because many people worry that calling into question the purported supernatural implications of near-death experiences will undermine some of their valuable effects.

Heightened moral sensibility is chief among those effects supposedly put in danger by the view we are here advocating. But what is the connection between moral sensibility and acceptance of supernaturalism? What is the connection between, on the one hand, caring about others and being attentive to their needs and, on the other, a rejection of physicalism and belief in the afterlife? It is not at all clear how the two are related. To simply insist that denying supernaturalism jeopardizes the salutary effects of near-death experiences just ducks the question. Why exactly do people who have had near-death experiences develop a greater sense of justice or compassion for others?

One possibility is that it is because people who have had near-death experiences take it that they have come into contact with an underlying truth about the universe and all of us in it. They have come to understand their place in the grand scheme of things. And this understanding has led them to change their orientation toward the world and others. Eben Alexander, for example, mentions love and a new and heightened awareness of its importance and centrality for everything. Certainly, a life truly steeped in unbounded love and lived in recognition of this fact is a morally admirable life indeed. And coming to a new and deep understanding of the role of love in one's life seems like the kind of experience that would transform one's behavior toward others in morally admirable ways. Perhaps this is all the more likely if this new understanding were the result of what one took to be a connection to a supernatural truth lying behind the world as one previously knew it.

But it is important to keep in mind that contact with the supernatural is not the only way in which these profound changes can come about. It is possible to come to a greater understanding of the universe and one's place in it through experiences understood in wholly physical terms. And it is certainly possible for this newfound understanding to transform one's moral character. The transformations we are talking about do not necessarily depend on the supernatural.

Consider the awesome spectacle of the Grand Canyon. Standing on the rim of the canyon is breathtaking—the beauty of the age-old sediment, eroded over the millennia by water rushing by as it still does today. The experience of viewing this scene is a reminder of the momentary nature of your existence, a reminder of the beauty the world has to offer. The cosmic scale of the whole thing—both in space and in time—is a reminder of the vastness of the universe, the limits of one's own experience. There is so much beauty to behold, and you will only ever behold so little of it. Try as you may, your capacity is limited; your time will run out. And the same holds for each person there on that rim with you. They also stand in awe; their time, too, will come. Whatever differences there may be between you, the similarities come to be all the more salient as you stand there, two creatures on the rim of a canyon on a floating rock in space, taking it all in together.

It's not hard to imagine—or recall—how an experience such as this can make a real difference in one's life, how it can make an impression that sticks. Maybe you once had the feeling of a shared plight with your fellow human beings and then lost it. But it can come back. The sight of a deep red sunset dusts it off. The smell of pine needles triggers something in your head. You feel a communion with others and the world around you, a feeling that began with that experience of seeing the awesome canyon for the first time. And it makes you think and act with a heightened awareness of your place in the world. You treat people better as a consequence. You no longer take your needs and interests to be any more important than theirs. In short, you become a better person.

The experience and transformation just described may strike some as dubious. How could viewing the Grand Canyon make one a better person? That's a good question. But the point is that it can. It does. People have long recognized the transformative power of coming into contact with the natural world. We recognize its power today. This is one reason that recovery programs are often set in the wilderness. It's why we take at-risk youth backpacking. No doubt, some people understand the natural world in relation to an underlying supernatural reality—as God's work. Others take it at face value. And yet all may be transformed by it. This goes to show that supernaturalism has no *monopoly* on transformative power. We can and do understand certain experiences in terms of the physical world, and this does not undermine their power to spur developments in our moral character.

So morally admirable transformation does not require acceptance of supernaturalism or even having had a near-death experience. Still, you might think we have missed the point. Seeing the Grand Canyon is one thing, but seeing heaven another altogether. And nothing we have said shows that a physicalist-friendly explanation of the latter experience wouldn't undermine its ability to effect moral change. It certainly seems plausible that explaining the experience of visiting one's deceased relatives in heaven in terms, say, of a drive to squelch one's terror at the recognition of one's own mortality would rob it of some of its apparent significance. To explain this experience in terms of the subject's psychology in this way would appear to extinguish its transformative spark.

This may be so. Maybe there is no way of coming to understand near-death experiences in wholly physical terms without robbing them of some of their transformative power. But this is simply beside the point when it comes to *explaining* what is happening when people have near-death experiences. The aim of explanation is to arrive at a true and accurate understanding of given phenomena. If the truth sets us free, all the better. If not, then we should seek other ways of becoming better people. It's not as if near-death experiences are the only (or even a widespread) source of moral transformation. There are myriad ways to become a better person. And it's not even as if all those who have near-death experiences are transformed by them. So even if making sense of one's near-death experience in physical terms would preclude the kind of moral transformation that some people undergo in light of these experiences, this would not have a huge effect on our overall moral landscape. It would not greatly reduce the numbers of morally good people. And though it may undermine what would otherwise have been a profound transformation for the better for some, it would not get in the way of their becoming morally better people in all sorts of other ways.

Finally, it is simply not clear that the kinds of transformations reported by those who have had near-death experiences would be undermined by a physicalist explanation of the phenomena. In fact, when we consider some of the details of the relevant transformations, they do not seem to require the backing of supernaturalism at all.

In his book, van Lommel offers helpful summaries of transformations typically associated with near-death experiences. One set of changes falls under the category "Self-Acceptance and a Changed Self-Image":

> The experience of transpersonal aspects during the NDE changes people's sense of who they really are. *Transpersonal* refers to those aspects of someone's consciousness that transcend the personal or the ego. This experience can be accompanied by a heightened sense of self-worth. Thanks to their changed self-image, people become less dependent on the approval of others, better at dealing with stress, and more adventurous, and they also take greater risks. It changes people's attitude toward their body and alerts them to new ways of

thinking. They are more likely to look at the bigger picture and are capable of forming more objective opinions, even at the risk of seeming aloof. And because they are more easily engrossed in things, they are less aware of their surroundings. Increased levels of curiosity coupled with a hunger for knowledge spark a particular interest in theological issues, philosophy, psychology, and natural sciences (especially quantum physics), although their education often fails to satisfy this need for deeper knowledge. They also develop a noticeably greater interest in physical and psychological processes and the possibility of (self-)healing.[6]

This is an impressive list. But what also jumps out is that none of these changes requires or even suggests the truth of supernaturalism. It is certainly possible that simple reflection on one's near-death experience and the fragility and brevity of life will issue in many, if not all, of these changes. Terrible tragedies and near-tragedies can focus the mind on what's important and meaningful without requiring us to give up a physicalist worldview. The conclusion that near-death experiences are best explained in physical terms need not result in the loss of any of the personal transformations on van Lommel's list.

The same may be said of a second set of remarkable changes in people who have had near-death experiences, which van Lommel offers under the rubric "Appreciating Life":

People who have a near-death experience seem positive of a new goal or new mission in life. They also appreciate the little

6. Van Lommel (2010: 52–53).

things in life, pay more attention to the here and now, and enjoy the moment. They are less likely, however, to allow themselves to be restricted by social convention. People are more confident of their ability to handle problems, more open to change, and less preoccupied with time and schedules. But even though they struggle with the concept of time, they do tend to honor appointments. They are better at putting things into perspective, they take an unbiased view of life, and they are quick to smile while at the same time more serious. Their increased respect for life also reveals itself in the greater appreciation of and interest in nature. They are now much more aware of seasonal change and like doors and windows to be open to admit fresh air. They also take more pleasure in classical or soothing music and are less tolerant of noise. NDErs attach less importance to status, money, and material possessions and distance themselves from the competitive elements in contemporary society.[7]

These are significant changes indeed. But these transformations in no way require a rejection of a physicalist interpretation of near-death experiences or belief in the existence of an afterlife. Consider the "new goal" or "mission" referred to in the first sentence. Though this may seem to require belief in an afterlife, it depends on what the mission is. And van Lommel leaves entirely open what it might be. For all he says, a person who has had a near-death experience may acquire a new mission of seizing every moment as if it were his or her last. So interpreted, the mission would not only fail to require belief in an afterlife but be in tension with it.

---

7. Van Lommel (2010: 54).

It is interesting to note in this context that many people report profound transformations—of strikingly similar sorts to the kinds described by van Lommel above—after having taken psychedelic drugs (either recreationally or as part of controlled medical experiments). Recently, there has been a renewed interest in the use of such drugs for the treatment of such conditions as post-traumatic stress disorder, depression, and death-anxiety in the terminally ill. Although of course the jury is still out, preliminary results are quite promising.[8] Michael Pollan writes:

> According to [Stephen] Ross [associate professor of psychiatry, NYU Medical School], cancer patients receiving just a single dose of psilocybin experienced immediate and dramatic reductions in anxiety and depression, improvements that were sustained for at least six months. . . .
>
> "I thought the first ten or twenty people were plants— that they must be faking it," Ross told me. "They were saying things like 'I understand love is the most powerful force on the planet,' or 'I had an encounter with my cancer, this black cloud of smoke.' People who had been palpably scared of death— they lost their fear. The fact that a drug given once can have such an effect for so long is an unprecedented finding. We have never had anything like it in the psychiatric field."[9]

We should reiterate that the results of this research are preliminary. But surely it is significant that so many research subjects have the sorts of transformations described by van Lommel—not in the context of near-death experiences, but rather due to the use

---

8. Some of this promise is described by Shroder (2014) and Pollan (2015).
9. Pollan (2015: 38).

of psychedelic drugs. These experiences are caused by a physical substance and its interaction with our brains. Of course, one might wish to interpret the experience as a physically caused opening up of the mind to nonphysical realms; we do not deny that this is a possibility. But what is salient and significant is that the kinds of profound transformations catalogued by van Lommel can indeed be induced by ingestion of a physical substance.

Once again, a supposed source of support for supernaturalism falls short. We are not warranted in accepting supernatural explanations of near-death experiences on the grounds that they are required for desirable transformations. These transformations do *not* depend on a supernatural explanation of these experiences. And furthermore, when it comes to the task of explaining near-death experiences, we should be aiming at the truth. To cite a supposed effect on our behavior and character is to appeal to the wrong kind of reason for accepting a certain kind of explanation. When we want to explain something, the question is what evidence there is for accepting a given interpretation of how things came about as they did, not how accepting a given interpretation is likely to transform us. A desire for something to be true is not evidence that it is true. Eben Alexander himself puts the point nicely: "Science is not concerned with what would be nice. It's concerned with what *is*."[10]

We wish to end this chapter with a bit of speculation about how a near-death experience could change one's moral orientation— and without invoking supernaturalism. Part of a near-death experience, typically at least, is an out-of-body experience. Here one might feel as if one is floating above one's own body, and it is as if

10. Alexander (2012): 35.

one can see one's own body from above. This change in perspective is at least a step toward the moral idea of stepping outside one's own perspective and taking a broader view of things, perhaps "putting oneself in another's shoes." Indeed, many moral theorists take something like this abstraction from one's own, narrow point of view to be a central feature of morality. Insofar as the out-of-body component of a near-death experience involves attaining a perspective that is untethered from one's body, it may be regarded as a first step toward taking what many regard as the distinctively moral perspective. It is at least possible that taking this step could lead to moral transformation. But it should be clear that taking a broader view of things or putting oneself in another's shoes does not require a belief in supernaturalism; taking a different *perspective* does not force us to believe in a different *reality* or *realm of being*.

Near-death experiences do not just issue in more attention to other people's interests; they can also result in less death anxiety and even more zest for life. Our contention is that these welcome changes in one's perspective do not depend on accepting supernaturalism. This point is driven home by the end-of-life reflections of an ardent physicalist, Oliver Sacks. Sadly, Sacks was diagnosed with terminal cancer while we were writing this book, and he has passed away. In a powerful article in the *New York Times*, he writes of the effect on him of receiving his diagnosis. Central among the changes he describes is having a more detached perspective: "Over the last few days, I have been able to see my life as from a great altitude, as a sort of landscape, and with a deepening sense of the connection of all its parts."[11] Sacks's enhanced understanding of the

---

11. Sacks (2015) ends his poignant and beautiful op-ed as follows:

> I have been increasingly conscious, for the last 10 years or so, of deaths among my contemporaries. My generation is on the way out, and each death I have felt as an

interconnection of all the parts of his life, and his corresponding sense of serenity, is achieved by a detachment from his particular perspective as an individual. He even describes this in terms similar to those used to describe an out-of-body experience, when he speaks of attaining a view "from a great altitude." But attaining this new perspective, and thus achieving this kind of illumination, clearly does not require supernaturalism. That is a view Sacks rejected until the end. Detachment from one's particular perspective may be facilitated by an encounter with death, and it may even require a leap of imagination, but it manifestly does not depend on the rejection of physicalism.

abruption, a tearing away of part of myself. There will be no one like us when we are gone, but then there is no one like anyone else, ever. When people die, they cannot be replaced. They leave holes that cannot be filled, for it is the fate—the genetic and neural fate—of every human being to be a unique individual, to find his own path, to live his own life, to die his own death.

I cannot pretend I am without fear. But my predominant feeling is one of gratitude. I have loved and been loved; I have been given much and I have given something in return; I have read and traveled and thought and written. I have had an intercourse with the world, the special intercourse of writers and readers.

Above all, I have been a sentient being, a thinking animal, on this beautiful planet, and that in itself has been an enormous privilege and adventure.

Chapter 11

A Strategy for Explaining Near-Death Experiences

It often seems as if certain phenomena must be "paranormal" and therefore outside the reach of science. When we scratch below the surface, however, alternative explanations emerge—explanations that are well within the physicalist framework of science. We have contended in this book that near-death experiences appear to fit this pattern.

In this chapter, we outline a preliminary strategy for explaining near-death experiences in physical terms. Our aim is not to present a complete explanation of any given near-death experience. We do not pretend to know all of the relevant facts in enough detail to be able to offer such an explanation. Rather, our aim is to provide what might be called a *preliminary blueprint* or *strategy* for generating candidate explanations for empirical scrutiny. We aim to demonstrate how one might go about constructing a multi-factor physical explanation of near-death experiences that is sensitive to all of their complexity, wonder, and transformational power.

Our proposed strategy may be put in terms of three steps:

1. Explain how the subject acquired the information relevant to some aspect of his or her experience;
2. Explain why this particular content would be included in a near-death experience; and
3. Explain when the experience took place.

This strategy fits our approach in earlier chapters. Consider, for example, the case of Pam Reynolds. We suggested that auditory sensations recording a conversation between the medical staff might have registered in her brain during her surgery without her being conscious of these sounds at the time. This is the first step. The second step would be to explain why Pam Reynolds would have the experience of hearing this conversation during her surgery. Here we might appeal to something like terror management theory. Perhaps it would allay some of her anxiety about her brain surgery to experience herself as consciously aware of what was happening in the operating room. Finally, and this is the third step, we suggested that Pam Reynolds's conscious experience of hearing the conversation between the medical staff may have occurred after the time at which it seemed to her that she had the experience. The unconsciously registered auditory sensations may have come together some time after her surgery, to produce a conscious auditory experience of hearing the conversation. Though it seemed to her as if this experience occurred during her surgery, it may, in fact, have been caused by processes that occurred later, perhaps while she was coming back to full conscious awareness and waking up from the anesthesia.

It is important to note that though our proposed strategy makes room for the kind of physicalist-friendly explanations we favor, it does not rule out supernaturalist explanations of near-death experiences. For instance, the first step in our strategy, explaining how the subject of the near-death experience acquired the information relevant to the particular aspect we seek to explain, does not require that this information be acquired at some other time than the experience itself. So our strategy does not rule out the possibility that one could perceive the way the world actually is during a near-death episode, for example, by actually hearing a conversation while one's brain is not functioning at all, or not in a manner capable of supporting conscious auditory experience. Thus, we are not dismissing out of hand explanations others have found compelling. What is crucial to us is not to rule out supernaturalist explanations of near-death experiences but rather to make room for explanations in wholly physical terms. We have already presented what we take to be excellent reasons to think that physicalist explanations are better than supernaturalist explanations. Our aim now is to present a strategy for developing what we take to be the superior kind of explanation of near-death experiences.

A second point is also worth highlighting. The first step in our proposed strategy does not require that the information relevant to some aspect of the experience be acquired at some time before or during the actual experience. It also allows that this information may have been acquired *after* the experience. We considered something like this when we suggested, as one of several possible explanations for how the man with the missing dentures came to know that they were in the drawer, that he might have seen other patients' dentures being placed in drawers during his stay in the hospital after his resuscitation. The second step in our proposed strategy, then, would be to explain how he came to experience seeing the nurse place his dentures in the drawer. This may seem impossible, given the claim that he came to know that his dentures were in the drawer only after he had had his near-death experience. But we do not think this is impossible. People can acquire false memories, including false memories of details of past experiences. The explanation of how the man with the missing dentures came to experience seeing what the nurse did with his dentures may involve, among other things, his coming to falsely remember this as being a part of the near-death experience he had while undergoing CPR. Perhaps this false memory was motivated, in part, by a desire to solve the puzzle of his missing dentures. No matter how he came to acquire it, this false memory might become a part of the near-death experience he later remembers having and reports to others.

While this may seem far-fetched, we do not think that it is. Indeed, we think that consideration of a near-death experience that has captured considerable attention, the case of Colton Burpo, provides support for this and other aspects of the three-step strategy we have just outlined. Before discussing that case, however, we will briefly consider a famous case of apparent reincarnation, the case of James Leininger.[1] The Leininger case highlights some lessons that will be useful in thinking through Colton Burpo's near-death experience.

At the age of two, James Leininger began having frightening dreams that caused him to yell out, kick, and claw in his sleep.

---

1. For some representative coverage of the Leininger case, see Brennan (2009), Telegraph (2009), ABC News (2005), and Milligan (2004). The Leiningers have told their story all over the world and have appeared on numerous TV and radio shows including *Coast to Coast AM, Larry King Live, Good Morning America,* and *Fox and Friends.* The 2009 news stories and appearances coincide with the release of their book *Soul Survivor.* A recent *Fox and Friends* interview (aired October 28, 2013) featuring a much older James may be seen at the following URL: http://foxnewsinsider.com/2013/10/28/amazing-leininger-family-believes-son-james-wwii-pilot-james-huston-reincarnated. (The writing of this section on James Leininger owes a great deal to the research of Heinrik Hellwig.)

He would dream that he crashed in an airplane and was trying to escape. The details James shared about these dreams were astonishing. He demonstrated encyclopedic knowledge about World War II aircraft, and he recounted facts about a particular aircraft carrier, the *Natoma Bay*, and its flight crew, including naming a fellow pilot, Jack Larsen. James's fascination with these aircraft and events in the war was not confined to his dreams. He drew pictures, acted out scenes, and talked frankly and in great detail about the battle of Iwo Jima. He signed his pictures "James 3" because he was, in his words, "the third James."[2]

As James's parents began looking more deeply into their son's behavior and interest in the war, they noticed some astonishing coincidences. The details young James was relating matched historical fact. The *Natoma Bay* was an actual ship that fought in the battle of Iwo Jima. The names of people James talked about in his dreams were members of the flight crew on that ship. One of the pilots shot down had been named James. They began to suspect that their son might be the reincarnation of that pilot. With the help of Carol Bowman, a therapist known for her work on cases of reincarnation in young children, the Leiningers began to listen more sympathetically to their son. What they saw and heard in his words and behavior only further supported their suspicion. They became convinced, even if reluctantly in the case of James's father, that their son had lived a past life as James Huston, who was shot down in his plane over the Pacific during the battle of Iwo Jima.

This brief synopsis should be enough to give you a sense of why many have thought that the best explanation of James Leininger's dreams and behavior is that his is a case of reincarnation. This was a very young child demonstrating a fascination

2. Bowman (2010: 56) and Milligan (2004).

with certain kinds of aircraft and apparent knowledge of particular historical people and events in ways that suggested that he was personally and deeply attached to them. It will come as no surprise, we're sure, to learn that we are not convinced by the reincarnation hypothesis. It may be one possible explanation of things, but it is not an especially compelling explanation. This becomes clear once we take account of the full range of factors, including both details of James Leininger's life and the overall explanatory context.

First of all, a young boy's fascination with airplanes, even airplanes of a very particular make and model, should strike no one as remarkable in the least. In general, and as every parent knows, young children become obsessed with people, places, and things. Often these obsessions are easily explained by exposure to the person or thing. For example, it is not uncommon for a young child to become fixated on a character from a book or movie after he or she has read or watched it—again and again and again. But sometimes these obsessions seem to come from out of the blue. It would be rash, however, to claim that just because one cannot understand how a child came to be fixated on something, the explanation must be supernatural. It is better to think, instead, that one has simply missed or forgotten something relevant to explaining how the obsession began.

James Leininger's obsession with World War II airplanes may strike one as an outlier, perhaps, because it is coupled with extensive knowledge of facts about both the aircraft and historical people and events. But before we jump to the conclusion that the best way to explain what is going on here is to appeal to reincarnation, we should consider whether there are any promising avenues for explaining things by more conventional means. And there do seem to be some possibilities here.

Consider the fact that the Leiningers visited a flight museum when James was eighteen months old. During their visit, James walked around the very type of plane he later claimed to have been flying in his dreams, Corsairs. His parents have even claimed that he looked like he was conducting a flight check.[3] Noting this detail about James's life seems like a promising start for providing an explanation of his obsession with World War II aircraft, and his fixation on Corsairs in particular. By all accounts James was a very bright child. It is possible that a bright 18-month-old could absorb many details at the air museum, especially in conjunction with listening to his parents or tour guides or others discussing the displays. James may not have had the ability at 18 months to put these experiences into words and express them verbally, but many of the details might have *registered*. Later, when he developed the capacity to verbalize, James would have been able to express the information that had registered earlier. We can thus begin to make sense of why James Leininger was so obsessed with World War II aircraft and how he knew what seemed like an uncanny amount about them. And we can do so without appealing to anything supernatural.

Even if James Leininger's detailed knowledge of World War II aircraft were to be explained in this way, it may still seem as though the case presents a serious puzzle. How could James have known the names of the people he recounted as being there with him in his dreams? We are not at all sure how the explanation of these facts is supposed to go. But it does seem possible to explain them without invoking reincarnation. Perhaps James heard these names somewhere. Maybe they were mentioned in a television program he saw (the Leiningers admit to conversing with James about a television

---

3. Leininger and Leininger (2009: 114).

program on World War II aircraft on the History Channel) or a book or museum exhibit panel someone (perhaps his parents) read to him. Perhaps his parents mentioned them in conversations at home. Children are incredibly perceptive, and their memories often outstrip those of adults. Given that he was obsessed with World War II aircraft, it would be no surprise if James soaked up tidbits of information that went in one ear and out the other for those less intent on the subject.

Once his parents and other adults began to ask him about these things, it would not be surprising if James's interest in them increased. He may even have begun to seek out more information about these matters out of a desire to please his parents and other authority figures, such as Carol Bowman and Jim Tucker (another expert on reincarnation), who interviewed him in relation to the possibility that he was reincarnated. The presence and interests of these people would be powerful influences on a young person at an age when the desire to please an adult authority figure is great. So there might have been a symbiosis between young James's desire to please and his parents' and certain researchers' beliefs and prior assumptions about what was happening in his case. This could have led to James's telling a story that suggested he was reincarnated, when in fact he was not.

Indeed, it would not be surprising if James Leininger came to "remember" events from a past life due to repeated, and possibly leading, questioning from his parents and other adults. It is a commonplace that the testimony of children is quite suggestible, and three- and four-year-olds have been shown to be more suggestible than even five- and six-year-olds. Among the factors thought to explain the phenomenon of suggestibility in children are, first, that their memories are not as firmly implanted as those in older people and, second, that the mechanisms for protecting and monitoring memories against suggestive intrusions are not as robust in young children as in older people. If James was repeatedly questioned about his dreams and his claims about a past life, especially by people who were themselves of the opinion that he had a past life, it would be consistent with psychologists' understanding of how memory works in young children to suppose that he came to falsely remember and report various facts and events.[4]

The psychologist Elizabeth Loftus describes the following three factors as involved in the formation of false memories: (1) social pressure to remember, (2) imagining an event as an aid to remembering it, and (3) being encouraged not to think about whether what one is remembering as happening actually happened. "False memories are constructed by combining actual memories with the content of suggestions received from others. During the process, individuals may forget the source of the information."[5] The three factors cited by Loftus may well have been present in James Leininger's case. Thus, it seems plausible that his apparent memories of his purported past life may have been false and, at the same time, eminently explainable in terms of psychological theories that have nothing to do with reincarnation.

All told, the explanation for why James Leininger had the dreams, told the stories, played the games, and said the things he did may be a *combination* of various factors. These factors may include chance coincidences, past events, normal childhood tendencies, and even suggestions and projections on the part of the adults involved in the case.

---

4. On the suggestibility of children, see Warren et al. (1991), Ceci and Bruck (1993), and Ceci and Huffman (1997).
5. Loftus (1997). For more on the ubiquity of false memories, see Chabris and Simons (2014).

The bottom line, as we see it, is that we should exhaust all possible avenues for explaining a given phenomenon by means of factors that fit into our usual, scientific worldview before we begin to take seriously factors, such as an immaterial consciousness or reincarnation, that do not fit into this paradigm. We have already noted one reason for this. There is a special problem that comes with appealing to an alternative paradigm: we have to make sense of how the elements of the alternative framework fit with the elements of our usual framework. For example, we have noted that if the supernaturalist thinks that consciousness is nonlocalized and immaterial, then he needs to help us to make sense of how the mind can interact with and control the physical body. A second reason for privileging the scientific paradigm is that this paradigm is widely shared and well understood, certainly more so than the supernatural alternative. So we not only need to make sense of how the elements of the two frameworks interact, but we also need to make sense of the alternative framework itself. What is an immaterial consciousness supposed to be like? How do we make sense of a nonlocalized consciousness in space and time?

Similar issues arise with respect to reincarnation. While the notion of reincarnation may seem to make sense of James Leininger's case, it is not at all clear that we can adequately grasp what the phenomenon of reincarnation is supposed to be like in detail or how it is supposed to fit into our wider understanding of our lives and ourselves. Here are some of the questions that come to mind when we think hard about reincarnation. Why, in a given case, was this person reincarnated as this other person? Is there some connection between past selves and future selves? Why are some people reincarnated and not others? Or, if one thinks that reincarnation is widespread, why do some people act as though they have been reincarnated and others do not? On a more personal note, why do most of us not feel as if we were reincarnated? What happens to people as they are waiting between lives? These are just a few obvious questions that cry out for answers if we take seriously the possibility that James Leininger is a reincarnated fighter pilot. It is not easy to see what the answers to these questions are. Nor is it easy to see where the questions stop. So it is not at all clear that appealing to reincarnation in this case makes any real explanatory progress. If anything, it seems to kick the explanatory can down the road, so to speak. In light of the possibility of explaining this case by appealing to more familiar factors, this seems like the intellectually responsible thing to do. This is not to say that there is no possible way of making sense of reincarnation no matter how hard we try, or that there are no intellectually rigorous attempts to do so. Reincarnation is an ancient doctrine with ties to belief systems around the globe. What we are saying is that the doctrine of reincarnation raises some deep puzzles, and in the absence of answers to these puzzles it is not clear that invoking it to explain a mysterious situation such as Leininger's makes any clear explanatory progress.

We stress that we do not take ourselves to have provided an adequate explanation of the Leininger case. That is not our aim. What we hope to have done is to have shown how the approach we have been sketching in this book may be applied to this case. Once we sweep aside unreasonable views about explanation and take care not to leap to unjustified conclusions, we can see how the Leininger case could, in principle, be given a physical explanation. Now we will consider how some of the lessons of the Leininger case might be applied to the explanation of near-death experiences. Our focus will be on the near-death experience of a young boy, Colton Burpo, as told by his father in *Heaven Is for Real*. This near-death experience has many commonalities with the Leininger case and also with the near-death experiences we have discussed in earlier chapters of this book. For this reason, it should be especially useful as we attempt to illustrate the overall strategy for explaining near-death experiences introduced at the beginning of this chapter.

Colton Burpo fell ill just a couple months shy of his fourth birthday and on the eve of a family vacation. After a day's reprieve during which the family headed out of town, things got worse. A few days later, his parents realized Colton did not have the flu, as they had thought, and took him to the hospital in the town they were visiting. It became clear that whatever was ailing Colton could not be treated in that small hospital, so they took him to a larger one closer to home. There he was diagnosed with a burst appendix and underwent surgery, twice.

After a miraculous recovery, Colton and his family finally returned home. But soon Colton began recollecting an experience he had had during his first surgery. Colton had visited heaven. He had personally met Jesus, God, and the Holy Spirit. He had met his deceased great-grandfather, Pop, and a sister who had never been born due to miscarriage. He saw angels, John the Baptist, and a rainbow horse. He even saw his parents in the hospital at the time of his surgery, his father in one room praying alone and his mother in another room praying and talking on the phone.

Colton Burpo had a "near-death episode."[6] He had emergency surgery for a life-threatening condition. And it seemed to him, after he recovered, as if he had had a very rich experience while he

---

6. We borrow this term from Holden (2009: 185) who defines a near-death episode as a *"physical situation* in which a person survives an actual or perceived close brush with death—typically, an acute medical crisis involving actual or threatened serious physical injury or illness."

was undergoing surgery. Moreover, this experience included many aspects typical of near-death experiences, such as an out-of-body experience, seeing deceased relatives, and having pleasant feelings. These are familiar from the near-death experiences discussed in earlier chapters. Both Pam Reynolds and the man with the missing dentures had out-of-body experiences. Eben Alexander experienced pleasant feelings and saw deceased relatives. Like Colton Burpo, he also described his experience as a visit to heaven, though not in the same biblical language and in terms of the same biblical imagery used by the young boy. We have discussed these other near-death experiences in some detail, arguing that these features of them can be explained by reference to physical factors. Now we wish to do the same with respect to Colton Burpo's near-death experience. This discussion brings together the various threads presented in earlier parts of this book.

Our basic aim is to illustrate how it may be possible to explain Colton Burpo's near-death experience without appeal to either tenet of supernaturalism—the existence of a nonphysical realm or nonphysical minds. We aim to show that a multi-factor physical explanation of his experience may be possible.

Let us begin by noting three striking parallels between the case for reincarnation made on the basis of James Leininger's behavior and the case for the reality of heaven made on the basis of Colton Burpo's testimony about what he experienced while undergoing surgery. First, both of these cases involve young children, and as a consequence, they are presented as especially surprising and authoritative. How could a young child come up with such information all on his own? It seems inexplicable that a two-year-old could relate such detailed information about a particular World War II battle, as James Leininger did, or that a four-year-old could have known what his great-grandfather looked like as a young man or that his mother had an unsuccessful pregnancy, as Colton Burpo did.[7] Furthermore, not only did these boys seem to get the details correct in their accounts, but they also appear to be unimpeachable sources. Todd Burpo is very explicit about this. He describes the "childlike humility" of children his son's age at the time of his near-death experience in terms of a "lack of an agenda."[8] The lesson he takes from this is that one who wants to get to heaven must somehow reacquire the "intellectual honesty" of a child.[9] Colton's character is not only beyond human reproach but also so pure as to be worthy of divine reward.

We have already discussed appeals to authority. In Chapter 5, we argued that Alexander's and van Lommel's appeals to their authority as physicians should not be taken to support their claims about the implications of near-death experiences for issues such as the relationship between the mind and the body and the existence of an afterlife. The special authority of physicians is relevant to medical, not metaphysical, questions. In our discussion of the Leininger case, we mentioned some specific reasons to adopt a healthy skepticism regarding the claims made by young children. They are not very good at distinguishing fact from fiction, and given their desire to please authority figures, such as their parents, there is a real danger that children can be led into stating as fact something that was suggested to them by another but did not actually happen.

Todd Burpo is keenly aware of the possibility that he might have led Colton into relating things that were not originally

---

7. Though Colton Burpo was three at the time of his near-death experience, he was four by the time he began telling his parents about it.
8. Burpo (2010: 74).
9. Burpo (2010: 75).

represented in his experience. And though he repeatedly mentions how he tried to avoid initiating conversations about Colton's near-death experience and asking leading questions, it is evident from Todd's own telling of the story that he was not able to stick with this resolve. For example, one night while reading Colton his nightly bedtime story from their book of Bible stories for children, Todd is prompted by a picture of Solomon's throne to ask his son, "Hey, Colton, when you were in heaven, did you ever see God's throne?" In response, Colton asks what a throne is and, once he has the one in the picture book pointed out to him, replies, "Oh, yeah! I saw it a bunch of times!"[10]

One natural explanation of this exchange is that Colton Burpo was prompted to claim that he saw God's throne because his dad asked him if he did. Todd Burpo thinks otherwise, however, because of the details his son provides about what he saw. Not only did Colton Burpo come out on his own with the claim that Jesus was sitting next to his Father, but when asked which side Jesus was on, he got it right. "There's no way a four-year-old knows that."[11] Unless, of course, like Colton Burpo, his father is a pastor, he attends church every Sunday, reads Bible stories every night before bed, and is surrounded by a community of worshippers just like him. Just as James Leininger could have absorbed details about World War II from a museum visit and television programs, a smart four-year-old, such as Colton Burpo, could have absorbed these kinds of details related in the Bible from the many, many hours he had already spent learning, directly or indirectly, about what the Bible says. When prompted by his father, a church pastor, to talk about these matters, it is not wild to speculate that Colton

10. Burpo (2010: 100).
11. Burpo (2010: 100–101).

Burpo pieced together (likely unconsciously) details from what he had absorbed over the years and related them as part of the experience his father had become so interested in. When asked which side Jesus was on, Colton got it right; but we should keep in mind that he had a 50/50 chance of being right! And if Colton's congregation recited the Apostles' Creed every Sunday, then he would have been very familiar with this detail.[12]

A second important parallel between these cases is that they involve very interested family members and experts (with prior orientations favorable to supernaturalism) who repeatedly question these children about what they apparently saw and heard. In James Leininger's case, he was questioned not only by his parents but also by experts on reincarnation who were antecedently strongly inclined to take seriously the possibility of reincarnation. In Colton Burpo's case, his parents, and especially his father (a deeply religious man), questioned him over a couple of years about his experience.

The details of Colton Burpo's near-death experience trickled out in small bits, with him sometimes speaking about it unprompted and other times responding to inquiries by his father, who was eager to learn more. On one occasion, after Todd Burpo mentioned to his own mother that his son had reported seeing Todd's deceased grandfather, Colton's grandmother drove into town to visit with him the very next day. Not only was Colton receiving special attention from his parents because of the experience he was relating about going to heaven during his surgery, but now his grandmother had made a spur-of-the-moment trip to his house to hear more. It is not a stretch to think that this sort of

---

12. We would like to thank an editor at Oxford University Press for this point about the Apostles' Creed.

encouragement might lead a young child to continue talking about an experience as if it were real, including more and more details as if they were a part of that experience. As we mentioned in connection with the Leininger case, it is even plausible that a young child might come to form false memories in this context. No matter how much Colton Burpo and his family may actually believe that he went to heaven, the fact that he was encouraged to continue talking about this reported experience by those he loves suggests a rival explanation for why he said some of the things he did.

A third parallel between the James Leininger and Colton Burpo cases is that both involve remarks about topics that the boys were introduced to beforehand. James Leininger was interested in planes and went to an air museum before he began having the nightmares. Colton Burpo is a pastor's son who reads Bible stories at bedtime and attends Sunday school and church at least once a week. It is not as if airplanes or heaven are surprising topics for these boys, respectively, to be interested in and talking about. In fact, given their ages, their interests, and the attention shown to them, we would expect them to talk about these very things a great deal. There is good reason to expect these boys to say things like what they are reported to have said, not on the basis of the reality of reincarnation or of heaven, but rather on the basis of what we know, in general, about children their age and about their life histories in particular. Recall that false memories are formed by combining actual memories with external suggestions. If Colton Burpo were already interested in and familiar with the accounts of heaven given in the Bible, it would not be surprising if he were to come to form false memories of having been there, informed by his antecedent knowledge of these accounts and the suggestions of those, like his father, interested in hearing more about this amazing experience.

It bears repeating that these boys' detailed knowledge of these subjects is not a clear challenge to the claims we are making. Young children are capable of absorbing a great deal of information, especially, as in Colton Burpo's case, when they are constantly surrounded by this information. Todd Burpo tells of his repeated astonishment that Colton's description of heaven matched the descriptions present in Scripture. But this is exactly what we should expect! After all, Colton Burpo is inundated with Scripture. He learns about what the Bible says at home, through his Bible bedtime stories, talking to his parents, hearing them talk to each other and to others, and at church, through services and Sunday school.

His extensive exposure to biblical stories and imagery may help to explain certain aspects of Colton Burpo's knowledge about the details of the Bible that surprise even his father. For example, Colton Burpo describes Jesus as having "markers," his word for stigmata. His father is surprised because he did not know if his son had ever seen a crucifix. "Catholic kids grow up with that image, but Protestant kids, especially young ones, just grow up with a general concept: 'Jesus died on the cross.' "[13] But it is highly likely that Colton Burpo had indeed seen a depiction of Jesus on the cross. For one thing, he reads Bible stories accompanied with pictures. These books are often quite graphic. And even if he never saw a depiction in his own books or those at his church, it is quite possible that he would have seen one outside a Catholic church or on a television program or in a hospital somewhere. The image of Jesus on the cross is ubiquitous, and especially so for those who are deeply Christian and rooted in deeply Christian communities. Todd Burpo's surprise is much more puzzling than Colton Burpo's familiarity with the image.

13. Burpo (2010: 67–68).

In considering these three parallels with the Leininger case, we have already begun to see that there are possible explanations of various aspects of Colton Burpo's near-death experience that do not require assuming that heaven is for real or that there is anything nonphysical at work. Now let us consider two more aspects of Colton Burpo's near-death experience—his out-of-body experience and his experience of seeing deceased relatives—and see how they might be explained in ways compatible with physicalism.

Colton Burpo reported having an out-of-body experience with apparently accurate contents. In this way, his near-death experience was similar to the near-death experiences of Pam Reynolds and the man with the missing dentures. But Colton Burpo's out-of-body experience is crucially different from these other two. The contents of Colton Burpo's out-of-body experience, as related by his father, are less detailed than the reported contents in these other cases. This is significant because, while the out-of-body experiences of Pam Reynolds and the man with the missing dentures represented unique ways the world was at a specific time, it is not so clear that Colton Burpo's out-of-body experience represented a unique way the world was at a given time. Here is how Todd Burpo recounts Colton's out-of-body experience:

> Colton said that he "went up out of" his body, that he had spoken with angels, and had sat in Jesus' lap. And the way we knew he wasn't making it up was that he was able to tell us what we were doing in another part of the hospital: "You were in a little room by yourself praying, and Mommy was in a different room and she was praying and talking on the phone."[14]

---

14. Burpo (2010: 61).

As is clear from this quotation, we do not have a very detailed report of this out-of-body experience. Yet Todd Burpo suggests that this aspect of his son's reported experience, above all else, supports the accuracy and truth of the contents of that experience: we can be sure that heaven is for real because Colton Burpo visited it at the same time that he experienced being outside of his own body and seeing his parents pray in separate rooms.

As opposed to reporting seeing unexpected or forgotten aspects of medical procedures, as Pam Reynolds and the man with the missing dentures did, Colton Burpo reported seeing his parents doing something that they presumably (and by their own accounting) do quite often. One would expect Colton's parents to be praying while he was in surgery. The striking fact here, according to Todd Burpo, is that Colton reported him, Todd, as having been in a small room alone while he was praying. This moment was very significant for Todd because he was getting angry at God. But the details related by Colton do not show that he actually saw his father in this significant moment. Rather, it seems that the most we can say based on the little description we are given out of Colton's mouth is that he reported seeing his parents doing something it would be reasonable for him to expect them to be doing at the time of his surgery. Granted, he reported seeing them doing it separately, but this need not be taken to support the claim that he actually saw his father in this fraught moment. Perhaps the Burpos do not always pray together or even normally pray separately. We are not told much about their prayer habits. Perhaps Colton had a particular memory or set of memories of his parents praying separately from other times in his life, and this memory, together with other factors, such as a desire to continue to please his parents by talking about his near-death experience, prompted him to relate this scene as if he had seen it during his surgery. From what we have to work with in the story as related by Colton's father, we cannot rule out this possibility.

Our possible explanation of Colton Burpo's out-of-body experience here fits the three-step strategy we introduced at the beginning of this chapter. First, we have suggested that it is plausible that he had seen his parents praying many times before his surgery. It is even plausible that he had witnessed them praying separately before. So we have the first step: an explanation of how the information forming this part of the content of his experience might have been acquired. For the second step, we might appeal to something like terror management theory. It is plausible that a visual representation of his parents praying would help to relieve some of Colton Burpo's anxiety about his severe illness and surgery. Whether or not he was consciously aware of it, it is not implausible that Colton Burpo knew something grave and distressing was happening to him. Perhaps the experience of seeing his parents pray for him would not only make sense to him at this moment (that seems like just the thing he would expect them to be doing during his surgery), but also serve a soothing function for him. This would explain why this representation was a part of his near-death experience. Finally, the third step: it may be that Colton Burpo had the experience of seeing his parents pray, not while he was undergoing surgery, but rather during the time he was coming back into consciousness. Perhaps he experienced seeing this soothing representation later than it seemed to him that he experienced it. And when he reported seeing his parents pray while undergoing surgery, he was mistaken about the timing of his own experience.

But even if we might be able to explain Colton Burpo's out-of-body experience in physical terms, can we explain his experience of seeing deceased relatives without appealing to the supernatural? Colton reported spending time both with a great-grandfather he did not know and a sister who had never been born and about whom he had never been told by his parents. These aspects of his near-death experience seem beyond the reaches of a physicalist-friendly explanation. His mention of his unborn sister, in particular, seems to pose an especially difficult challenge to the strategy we favor.

Yet again we need to be careful here. It is not at all clear that Colton came out with these details in their fullness on his own and without the (however unintentional) prompting of his parents and others. Consider, for instance, the scene in which Colton relates meeting his unborn sister, as told by his father:

> I heard Colton's footstep padding up the hallway and caught a glimpse of him circling the couch, where he planted himself directly in front of Sonja [his mother].
>
> "Mommy, I have two sisters," Colton said.
>
> I put down my pen. Sonja didn't. She kept on working.
>
> Colton repeated himself. "Mommy, I have two sisters."
>
> Sonja looked up from her paperwork and shook her head slightly.
>
> "No, you have your sister, Cassie, and ... do you mean your cousin, Traci?"
>
> "No." Colton clipped off the word adamantly. "I have two *sisters*. You had a baby die in your tummy, didn't you?"
>
> At that moment, time stopped in the Burpo household, and Sonja's eyes grew wide. Just a few seconds before, Colton had been trying unsuccessfully to get his mom to listen to him. Now, even from the kitchen table, I could see that he had her undivided attention.

"Who told you I had a baby die in my tummy?" Sonja said, her tone serious.

"She did, Mommy. She said she died in your tummy."[15]

Even supposing that this is how the conversation went, a few things jump out. First, Colton was clearly rewarded by making the bold statements he did. As his father puts it, he was trying to get his mom's attention and he finally succeeded. This isn't to say that Colton was just making stuff up to shock his mother into paying attention to him. But it is illuminating to put this scene in the context of Colton's life since he began telling his parents about his near-death experience. It does not seem out of the question that he learned a certain behavioral pattern, roughly, that when he made comments of a certain nature and with a certain directness, his parents snapped to attention. Like when his grandmother paid an impromptu visit after he talked about meeting his great-grandfather in heaven, Colton certainly got a response from his parents when he told them about meeting his deceased sister in heaven.

Why should we rule out the possibility that Colton Burpo coincidentally said something here that struck a chord with his mother and then ran with it because he could see how much it interested her? We need not assume that he came up with this subject out of the blue. Even though his father insists that they never talked to their son about this topic, he does admit that they talked to his older sister about it and that it was a significant event in their lives. Todd and Sonja Burpo likely talked to each other about it at times. It is not out of the question that Colton heard about the miscarriage from his sister or overheard his parents talking about it sometime when they were unaware that he was listening. So there

15. Burpo (2010: 94).

are possible explanations for Colton knowing about his mother's miscarriage other than his having met his unborn sister in heaven, as his parents would like to believe.

As with his out-of-body experience, our possible explanation for Colton Burpo's experience of seeing his deceased sister in heaven also fits the three-step strategy introduced at the beginning of this chapter. First, we have suggested that Colton Burpo might have heard about his mother's miscarriage from his sister or overhead his parents talking about it when they were not aware that he was listening. So he could have known about his deceased sister before his surgery. Second, it seems plausible that the representation of another child of his parents who had already died might be significant, and possibly even comforting, at a moment when he was himself brushing with death. Colton Burpo might have experienced seeing his deceased sister because it gave him comfort to know that he was not the first child his parents would have lost and that he would not be isolated from his entire nuclear family after passing on. Alternatively, we have suggested that Colton Burpo might have reported this aspect of his experience in part because he wanted to please his parents and it got their attention. A false memory of seeing his deceased sister might be explained, in part, by a desire to please his parents. Finally, it might have been the case that Colton Burpo experienced seeing his sister, not while undergoing surgery, but at some later time. As with his out-of-body experience, this may have been as he was coming back into full conscious awareness. Or perhaps he came to falsely remember experiencing this at some time much later.

In taking seriously the possibility that Colton Burpo's near-death experience is to be explained, at least in part, in terms of a false memory, are we going back on our repeated earlier claims that we are willing to take people's reports of their near-death experiences at face value? How is positing a false memory of an experience consonant with accepting that the experience was real?

This is a good question, but we have not gone back on our earlier claims. We are not claiming that a false memory might explain the entirety of Colton Burpo's near-death experience. Rather, we are claiming that a false memory might explain one specific aspect of that experience. So we are not calling into question the claim that Colton Burpo actually had a near-death experience, but only the claim that, at the time he had this experience, it contained all of the details he later remembered it to have contained. Moreover, we are not claiming that Colton Burpo intentionally misled anyone about the nature of his near-death experience. False memories are just as real to the one remembering them as are memories based on actual experiences. From Colton Burpo's perspective, a false memory of seeing his deceased sister would be just as real a part of the experience he remembers having as any other aspect that was actually a part of his experience at the time he had it. Just as we can distinguish between the time at which it seemed to Colton Burpo that he had his near-death experience and the time at which the experience actually occurred, we can also distinguish between the specific contents it seems to Colton Burpo that his experience had *at the time he is reporting this experience* and the actual contents that experience had *when he experienced it*. In making this distinction, we make room for the possibility that certain aspects of the experience as he reported it are to be explained by appeal to false memories.

This leads us to make a related point about the context in which near-death experiences are recorded and reported to the general public. The details of Colton Burpo's case are complicated by the fact that his story is told by his father (with the help of a ghost writer). So we are not getting the details directly from the source. And this feature is more or less shared by every report of every near-death experience that comes to the public's attention. If the actual words relating the near-death experience are not given by a third party, as in the case of the man with the missing dentures, then they are disseminated to us, the general public, by those who have enough interest to publish or broadcast them.[16] It is, for example, because others took a keen interest in her story that we have such a detailed account and thorough discussion of Pam Reynolds's near-death experience.

In the case of children, there is good reason to think that their sense of what certain people want to hear may have an effect on the reports they give of their experiences. But why think this phenomenon holds only in the case of children? Even adults like to please others. It is not out of the question that one might (unintentionally and unconsciously) embellish the details of a given experience in order to fit the expectations of those interested in it. This possibility becomes all the more probable when those interested in the story have something of a stake in how it gets told and what it contains—even if this stake is just confirmation of their antecedent beliefs about the nature of reality. It is simply not clear that any case from the literature on near-death experiences is untainted by the expectations and assumptions of those interested in the implications of what is being said. And given this, it is unclear to what extent we can rule out various alternative explanatory possibilities.

Even if a particular detail in a report of a given near-death experience presents a seemingly insurmountable hurdle for those

---

16. The recent case of Alex Malarkey is a reminder that there are often powerful interests behind the dissemination of these stories. The details in Dean (2015) are striking in this regard.

wishing to explain that near-death experience in physical terms, the possibility may remain that this detail is the result, not of the experience that needs explaining, but rather of how that experience came to be recalled and reported. In other words, when we consider how to explain near-death experiences, we need to keep in mind that they are reported to us in specific contexts, and the proper explanation of some of the details in these reports may be due to factors having to do with their dissemination, and not with the near-death experiences themselves. Sometimes, that is, we may be searching for a way of explaining how someone came to experience something, when what really needs explaining is how they came to *report* experiencing that thing.

We are not accusing anyone of intentionally falsely reporting near-death experiences or of providing deliberately misleading reports of cases or deliberately asking leading questions. Rather, we are pointing out that the issues are complex and that care is needed in how we approach the task of explaining near-death experiences, especially given the nuances of the circumstances surrounding their dissemination. We are all human beings, and it is sensible to keep in mind our basic psychological tendencies. It is all the more sensible to do so when one of the proposed alternatives is to accept an explanation of what is going on in terms of supernatural phenomena that challenge our common-sense understanding of the way the world—including us—works.

Chapter 12

Confirmation Bias

We Believe What We Want to Believe

Why do people gravitate so powerfully to supernatural interpretations of near-death experiences? Why is the sort of explanation we have sketched (and, admittedly, it is only a sketch) not more visible in discussions of near-death experiences and attractive to those who think and write about them? It is a good idea, as we noted at the end of the previous chapter, to keep in mind that we are all human beings with certain psychological tendencies. This can help to explain why multi-factor physical explanations of near-death experiences are often ignored. Human beings sometimes look past the obvious. And usually we can understand why.

Let's begin with a nice story told by Stuart Vyse:

One autumn day, I took my then three-year-old son to the park. It was the first really cold day of the season, and he was wearing a pair of red mittens. I was wearing grey woolen gloves. As I pushed him on the swing, I asked, "What is the difference between mittens and gloves?" His answer was very confident: "Gloves don't have trains on them." He had made an inductive error. I had asked him to draw a conclusion, to extract a general rule regarding gloves and mittens. His job would have been simpler had the items in question differed in only one way but this was not the case. Our handwear was of contrasting colors, sizes, and shapes (fingers versus no fingers), and, most important, the backs of his mittens were embroidered with a train engine on each hand. Trains have always played an important role in my son's life; therefore, the lack of trains on my gloves was, by far, the most salient distinction to his young mind.[1]

Vyse continues:

. . . [M]ore sophisticated, adult scientific reasoning often involves induction. A test is devised, and based on its results, conclusions are drawn. Errors, like that committed by my son, are common when all the alternative conclusions have not been considered (fingers versus no fingers), and they are particularly likely when the scientist is committed to one answer. In this case, the investigator is apt to construct the test so as to validate his or her beliefs. This last example is known as confirmation bias, and it can be an obstacle to effective reasoning for scientist and nonscientist alike.[2]

When we exhibit confirmation bias, we fail to consider all relevant hypotheses or explanations. We tenaciously cling to our existing beliefs. We seek out information that accords with those beliefs, and we avoid information that is in conflict with them. We fit data to a theory that we are predisposed to like by noticing the

---

1. Vyse (1997: 119–120).
2. Vyse (1997: 120).

confirming hits but not the misses. Confirmation bias is, at bottom, a syndrome of tendencies that protects our already existing beliefs.

Human beings have a natural tendency toward confirmation bias. But it is surely magnified in contexts in which our protected beliefs are extremely important to us, perhaps because they give us a sense of comfort and security. And it is obvious that supernaturalism may be extraordinarily comforting. Consider, for example, the belief that one will be ushered into a beautiful, serene afterlife by one's deceased loved ones. This is a recurring theme in the supernaturalist interpretations of near-death experiences we have been discussing. It's a comforting thought, one that people might be naturally inclined to protect. Eben Alexander gives a stark description of why this may seem worthwhile. After writing that he had "grown up wanting to believe in God and Heaven and an afterlife," he continues:

> As a doctor who saw incredible physical and emotional suffering on a regular basis, the last thing I would have wanted to do was to deny anyone the comfort and hope that faith provided. In fact, I would have loved to have enjoyed some of it myself.[3]

Again, the basic tenets of terror management theory provide a good starting point for thinking about these matters and how to explain them. We are all afraid of death. Indeed, we usually take it that death is bad insofar as it deprives us of the good things of life. And those goods include, most centrally, connections to people we love and projects that are "bigger" than we are. How

---

3. Alexander (2012: 34, 35).

extraordinarily comforting it would be to know that death is only a transition to a place where we would *continue* to be conscious, engaged in the projects that provide meaning in our lives, and connected to the people we love. Although some links to people we care about would be broken, this would only be temporary, and we would immediately reconnect with loved ones whom we have lost. Moreover, the afterlife would have the over-arching structure of something significantly "greater" and "higher" than the mere human. It would be rooted in the divine.

Richard Posner nicely captures the way in which confirmation bias can combine with other psychological tendencies to prompt us to form communities of like-minded people. As he puts it, "Nothing is more reassuring, so far as the felt soundness of one's beliefs are concerned, than to find an intelligent, articulate person who shares them and who can make arguments and marshal evidence for them better than you yourself can and thus arm you to defend them better if challenged, as well as to still your own doubts."[4] But, as Posner also notes, surrounding oneself with those of like mind is not the best method for arriving at the truth. In order to do that, we ought to challenge our convictions and see how they hold up in the face of plausible disconfirming evidence. But that is just the point. Confirmation bias is not a tendency aimed at the truth so much as a tendency aimed at preserving one's current stock of commitments.

The fact that we are subject to confirmation bias is particularly salient in the context of discussing rival explanations of near-death experiences. It is on prominent display in some of the most influential discussions of them. One way of characterizing Eben

---

4. Posner (2003: 148).

Alexander's second book on the topic, *The Map of Heaven*, would be as one long exercise in confirmation bias. This book contains an extended exposition of the "benefits" to be secured by a belief in an afterlife, coupled with the claim that near-death experiences are the key to understanding everything. The benefits in question correspond to the seven chapters of the book: Knowledge, Meaning, Vision, Strength, Belonging, Joy, and Hope. Alexander's aim is to specify various reasons why his reader will want to confirm that heaven is for real, and he does so by giving his readers reason to seek and interpret evidence in a way that will confirm this proposition. He does not examine arguments against supernaturalism and offer reasons for why the view holds up in the face of them. Rather, he marshals only evidence in favor of the view he holds to be true and doubles down with appeals to the benefits of maintaining belief in it. This is something of a textbook example of many of the tendencies associated with confirmation bias.

But Alexander's books sell. People want to hear what he is saying. This is confirmed by the following anecdote. The son of one of the co-authors of this book, Ariel Fischer, was washing his car at his home in San Francisco, when an older neighbor approached him and started raving about Eben Alexander's *Proof of Heaven*. Ariel mentioned to the neighbor that his father was working on a book on the topic. She asked him to describe the book, and he did. The neighbor interrupted him before he could even finish, saying, "I will *never* read your father's book! *I want to believe Eben!*" But wanting something does not make it so. Alexander's books might strike a chord with millions of his readers, and yet the view he is out to convince us is True with a capital "T" might be False with a capital "F." His arguments are not well designed to help us to find out which is the case. They are meant to speak to the initiated, not get to the bottom of things.

Confirmation bias is nothing new. Its effects on sound argumentation were noted in this great passage from Francis Bacon in the 17th century:

The human understanding when it has once adopted an opinion (either as being the received opinion or as being agreeable to itself) draws all things else to support and agree with it. And though there be a greater number and weight of instances to be found on the other side, yet these it either neglects and despises, or else by some distinction sets aside and rejects, in order that by this great and pernicious predetermination the authority of its former conclusions may remain inviolate. And therefore it was a good answer that was made by one who, when they showed him hanging in a temple a picture of those who had paid their vows as having escaped shipwreck, and would have him say whether he did not now acknowledge the power of the gods—"Aye," asked he again, "but where are they painted that were drowned after their vows?" And such is the way of all superstition, whether in astrology, dreams, omens, divine judgments, or the like; wherein men, having a delight in such vanities, mark the events where they are fulfilled, but where they fail, though this happen much oftener, neglect and pass them by. But with far more subtlety does this mischief insinuate itself into philosophy and the sciences; in which the first conclusion colors and brings into conformity with itself all that come after, though far sounder and better. Besides, independently of that delight and vanity which I have described, it is the peculiar and perpetual error of the human intellect to be more moved and excited by affirmatives than by negatives; whereas it ought properly to hold itself indifferently disposed toward both alike. Indeed, in the establishment of any true axiom, the negative instance is the more forcible of the two.[5]

Bacon's story of the shipwreck nicely illustrates how we tend to gravitate toward confirming instances of cherished beliefs and ignore countervailing evidence. A story that makes a similar point was told to a colleague of ours by his mother. His mother's mother and sister were Christian Scientists who both died fairly young of breast cancer. During Christian Science ceremonies, there are testimonials where people stand up and say what they have been cured of. Our colleague's mother went to the Christian Science church in Boston and during the testimonial stood up and said that she was there to testify for the people who had prayed and died, who couldn't testify for themselves, like her mother and sister. She sat down and the next person to testify said, "They didn't pray right." Alas, human beings are *tenacious* in clinging to their cherished beliefs; it can get in the way, even, of compassion at another's loss.[6]

Perhaps the tendency to seek out confirmation of one's deeply held beliefs also helps to explain why hundreds of people have sent us email messages in connection with the Immortality Project, a $5,000,000 grant from the John Templeton Foundation that allowed us to fund more than thirty academic research projects on various topics, including near-death experiences. We were inundated with ardent, detailed, sincere messages from people who wanted to share their near-death experiences with us and relate what these experiences revealed about the true nature of reality. They wanted

5. Bacon (1620/2004: Part One, XLVI).
6. Pierce called confirmation bias "the method of tenacity": Pierce (1877/1997).

to let us in on the evidence for supernaturalism. But why did so many people reach out to *us*? There are many people thinking seriously about near-death experiences, and many of them, as we have seen, are outspoken and on the record in favor of supernaturalism. No doubt some of the people who wrote to us wished simply to advance the cause of scientific inquiry into a subject they cared deeply about. But it is also plausible that many felt a need to have their stories "validated" by experts. They wanted their interpretation to be confirmed by those running a major grant at a research university. Not only is there strength in numbers, but one's belief is more secure if it is validated by authoritative figures and institutions. Why else would *so many* people implore us to believe in the reality of what their near-death experiences seemed to show them—to believe that these were indeed proof of supernaturalism? By simply announcing that we were interested in learning more about these experiences and their implications, we seemed to have put out a call for testimonials. It seems likely that the driving force behind the efforts of many of those who reached out to us in such a heartfelt and sincere way was at some level a desire for validation of their belief in supernaturalism. We hope that those who have shared their fascinating and gripping stories with us are open to hearing what we have to say on the subject as well.

Consciously or not, we human beings employ various strategies to manage our fear of death. Confirmation bias is among them. But this strategy can lead us to ignore what is right in front of our eyes. In this chapter we have pointed out that this can help to explain why so many people are attracted to the supernatural interpretation of near-death experiences rather than the sort of multi-factor physical explanation we have identified as a serious contender. One of our main claims has been that physical explanations of near-death experiences are more likely to be true, all things considered, than the supernatural explanations preferred by many (if not most) who have publicly discussed the topic. In Chapter 13 we contend that such physical explanations are not only significantly more likely to be *true* than supernatural explanations, but that they are also capable of being deeply *attractive* and even deeply *inspiring*. We don't just want truth; we want beauty and inspiration as well.[7]

7. We are grateful to Heinrik Hellwig for his help in preparing this chapter.

Awe, Wonder, and Hope

Near-death experiences elicit deep awe and wonder, and they inspire great hope. One might worry that our project here, of explaining near-death experiences in wholly physical terms, would eliminate these features. The concern stems from the idea that if near-death experiences can be explained physically, this would threaten the meaningfulness of life, and with it, awe, wonder, and hope. But just as we argued (in Chapter 10) that we need not cede the territory of profound transformation or moral sensibility to those who invoke the supernatural in order to explain near-death experiences, we will argue here that we need not cede the territory of meaningfulness. That something can be explained physically does *not* imply that it cannot be seen as deeply meaningful, and as capable of inspiring awe, wonder, and hope.

The worry seems to rest on an outdated view about the relationship between scientific inquiry and affective responses, such as awe and wonder.[1] It assumes that these responses are stifled by scientific explanation. Wonder at the incomprehensible may prompt us to seek explanations, it is assumed; but these explanations dispel our awe by pulling the phenomena back down to earth. This may

---

1. For an excellent discussion of these issues, see Daston (2014).

have been the ancient or medieval view. But nowadays we recognize that the outputs of science—the very explanations prompted by our curiosity in the face of a lack of understanding—can themselves inspire feelings of awe and wonder. Science can touch our emotions by presenting us with elegant explanations.

Everyday experience confirms the power of the natural world to inspire deep feeling in us. It is evidently not the case that only experiences understood in supernatural terms may be powerful and wonderful. Think of the beauty of a sunset. More specifically, think, again, of a sunset over the Grand Canyon. To witness this daily event is to be taken aback by breathtaking beauty. A beautiful sunset can fill us with a sense of awe and wonder. But a sunset is a purely physical phenomenon; it is the result of the interaction of light waves with water vapor. We can be filled with awe and wonder even with the fact clearly in mind that what we are witnessing is a purely physical phenomenon. Think also of majestic mountain peaks, giant redwood trees, and the power of a thunderstorm. Awe and wonder are not only appropriate in reaction to the supernatural.

Contemplating the products of human design and coordination can even fill us with awe. Think, for instance, of stunning modern skyscrapers or the pyramids in Egypt. When contemplating the fact that human beings designed and built the pyramids, one can be filled with a sense of awe at this remarkable human accomplishment. And yet, the pyramids are manifestly physical phenomena and clearly the product of human toil. Indeed, it is partly the recognition of this human toil, unaided by machines, that inspires awe at the pyramids. Thus, awe and wonder need not even be a reaction to the natural, in a sense that requires separation from human intervention. And they need not be reactions to something bigger than oneself. Though many of the human creations that inspire awe and wonder are the work of many individuals together, this need not be the case. One can experience awe and wonder at one's own achievements. Think of baking a perfectly shaped loaf of bread, running a marathon, or painting a beautiful landscape. These are individual achievements one might not only take pride in but also appreciate with a sense of awe and wonder at what one has done. And this appreciation does not require regarding these as acts of supernatural origin in any sense.

Howard Wettstein, in a fascinating study of awe, offers a helpful catalogue of various kinds of contexts that do not necessarily imply anything supernatural but that typically invoke deep awe:

1. Cases of awe at natural grandeur; the feelings of an astronaut standing on the moon, or someone powerfully moved by the night sky at the top of a mountain, or a relevantly similar ocean experience.

2. Awe at human grandeur. There are examples that seem available to everyone, given a certain openness and sensitivity: awe at the power of people to find inner resources in horrible circumstances, awe at human goodness and caring. Other examples require artistic and/or intellectual sophistication: powerful responses to great art of all varieties, or to great achievements in science, mathematics, philosophy.

3. Awe at the birth of one's children. Perhaps this is a compound of or intermediate between 1 and 2.

4. Even more sophisticated, more rare, are other sorts of combinations of 1 and 2. For example, one at the top of the mountain, awestruck not only by the overwhelming beauty

and majesty of nature, but also by the fact that humans, constructed of the stuff of the mountain, can take such a thing
in, and indeed that they can feel awe at it.[2]

The key idea Wettstein expresses here is that these contexts are all
apt to produce deep reactions in human beings, and yet it is not
tempting to invoke supernatural objects or causes.

You might agree with all of this but still insist that there is
something fundamentally different about near-death experiences.
These are not artifacts, like the pyramids, or natural wonders, like
the Grand Canyon, but rather experiences with special characteristics. You might think that this sets them apart. Perhaps we can
appreciate with awe and wonder the beauty of the natural world
or the fruits of our own labor without regarding the objects of our
appreciation as supernatural in any way. But the same is not the
case with respect to near-death experiences.

Why should near-death experiences be regarded as special in
this way? One thought would be that it is due to the special phenomena characteristic of them. But there are other kinds of experiences characterized by similar phenomena. And these experiences
may be awe-inspiring and meaningful. So it does not seem that
the significance of near-death experiences is due to their special
characteristics. Moreover, these similar experiences are not ones
that we are tempted to understand in supernatural terms. And yet
they inspire awe and wonder. So it does not seem that the awe and
wonder engendered by near-death experiences is due even to the
combination of these special characteristics and understanding
them in supernatural terms.

2. Wettstein (2012: 30).

What are these similar experiences we are referring to? Consider the following description of an LSD trip:

I was in my late twenties when a friend and I took some LSD. I had tripped many times before but this acid was different. . . .

We noticed that we were talking to each other mentally through thoughts only, no verbal talk, tele-communicating. I thought in my head, "I want a beer," and he heard me and got me a beer; he thought, "Turn the music up" and I turned the music up. . . . It went on like this for some time.

Then I went to urinate, and in my urine stream was a video or movie of the past played back in reverse. Everything that had just happened in the room was coming out of me like watching a movie in my urine stream, playing in reverse. This totally blew my mind.

Then my eyes became a microscope, and I looked at my wrist and was able to see each individual cell breathing or respirating, like little factories with little puffs of gas shooting out of each cell, some blowing perfect smoke rings. My eyes were able to see inside each skin cell, and I saw that I was choking myself from the inside by smoking five packs of cigarettes a day and the debris was clogging my cells. At that second I quit smoking.

Then I left my body and hovered in the room above the whole scene, then found myself traveling through a tunnel of beautiful light into space and was filled with a feeling of total love and acceptance. The light was the most beautiful, warm and inviting light I ever felt. I heard a voice ask me if I wanted to go back to Earth and finish my life or . . . to go in to the beautiful love and light in the sky. In the love and light was every person that ever lived. Then my whole life flashed in my mind from birth to the present, with every detail that ever happened, every feeling and thought, visual and emotional was there in an instant. The voice told me that humans are "Love and Light.". . .

That day will live with me forever; I feel I was shown a side of life that most people can't even imagine. I feel a special connection to every day, that even the simple and mundane have such power and meaning.[3]

The experience described here has many of the hallmarks of a near-death experience—a life review, an out-of-body experience, traveling through a tunnel of light, pleasant feelings, deceased people—and the subject experienced it as powerfully meaningful. Yet it was induced by LSD, and he knew this, both as it was happening and when he reflected on it in the retelling. This man's awareness of the physical explanation of his experience does not seem to have diminished its significance for him in the least.[4]

This LSD case strongly suggests that it is *not* necessary to understand experiences very similar to near-death experiences in nonphysical or supernatural terms in order to appreciate them with awe and wonder. But perhaps this is too hasty. After all, there does seem to be a clear difference between the way subjects of near-death experiences relate to the contents of their experiences and the way in which it would be most reasonable for the subject of this LSD trip to relate to the contents of his experience. Eben Alexander, Colton Burpo, and many others who have near-death experiences take the contents of their experiences to be accurate. Even if it requires expanding their conception of what the world is like, these people take their experiences to represent reality.

3. Sacks (2012a: 101–102).
4. For many similar examples of profound spiritual experiences induced by psychedelic drugs, see Shroder (2014) and Pollan (2015).

However, it would seem unreasonable to both hold in mind the fact that one's experience is caused by a hallucinogen and also to take it to represent reality. The point of taking a hallucinogen is to check out of the real world for a while. Certainly we should not interpret this man's LSD trip as containing more than illusions. And we shouldn't take him to think that it contains anything different.

But then the sense in which this experience was meaningful to him *does* appear to be different in a crucial respect from how near-death experiences seem to be meaningful to those who have them. In the case of a near-death experience, some of its significance may be attributed to the fact that it is interpreted as putting the subject in touch with aspects of reality that would otherwise be closed to him. It is meaningful, at least in part, because it allows him to gain a perspective on this world that he would not otherwise have or because it reveals that heaven is for real. But the content of an LSD trip is not taken to present aspects of reality in this way. What one sees while tripping is not taken to be real, at least not after the moment has passed. The feelings one has while seeing the sights, or the connection one feels to others along the journey—these may be interpreted as very real emotions and connections. But the reality and importance of these feelings does not rest on the accuracy of the contents of the experience in the way that the significance of near-death experiences seems to rest on the purported reality of their contents.

So there is a genuine distinction to be made between the significance of near-death experiences and the significance of experiences of similar contents, such as the LSD trip described above. In particular, there is a real case to be made for the claim that the significance of near-death experiences may depend on a supernatural interpretation, even if the significance of an LSD trip with similar features does not. If the significance of a near-death experience depends, at least in part, on the accuracy of its contents, and if these contents are supernatural in nature, as heaven surely is, then perhaps the awe and wonder this experience provokes would be undermined by a wholly physical understanding of it.

This is a good challenge to our claim that a physicalist-friendly understanding of near-death experiences would not undermine their significance. But we think it can be overcome. This interpretation of the significance of near-death experiences takes the apparent accuracy of their contents to be central. This is the basis for distinguishing between them and the LSD trip, which is significant and meaningful even though it is not taken to represent reality. But there is a different way of making sense of the meaning of our experiences that does not emphasize accuracy in this way. And we think that this alternative conception of meaningfulness is both generally applicable to human experience and also helps to point to a shared manner in which both the near-death experiences and the LSD trip may be meaningful to their subjects.

Consider the difference between explanation and storytelling. These are two deep-seated aspects of our human nature, distinct ways in which we seek to come to grips with the world and ourselves in it. Storytelling is how we sort through the significance of what happens to us. Sometimes our stories are entertaining, and sometimes profound. Other times they are dry, factual, and to the point. Whatever they are like, the stories we tell help us to come to terms with our lives . . . and also our deaths. They help us to sort out this experience we call living. And they do so by placing events into emotionally recognizable patterns. We feel the pull of narratives because they take us—both in body and in mind—through recognizable emotional landscapes. We feel the tension of drama, the crushing pain of tragedy, the comic release. This is the distinctive way in which stories make sense to us.

But we don't just tell stories. We explore the world, including ourselves, in an effort to understand the way things work. We use the tools of science to discover what is out there—and in here—and how it all fits together. We observe, hypothesize, and test. We refine our vocabulary for describing what we find, and we constantly revise our explanations for why things are the way they are. In doing so, we are searching for the truth. Unlike our drive to find meaning, our drive to explain is not satisfied by fictional representations. A good explanation touches the world, just as a good story touches the heart.

These two enterprises—storytelling and explanation—are uniquely and essentially human. When it comes to making sense of the world and ourselves, human nature is multi-faceted. We want to understand the way things work, and we want to grasp the meaning of it all. These two pursuits are not necessarily in tension. Sometimes they work in tandem. Placing an explanation in the context of a narrative can be a powerful way of getting the message out and making it comprehensible. It can help to communicate the deep significance of the events being explained, and to help people feel the importance of the topic. But other times these two sides of human curiosity are at odds. Dull explanations are unsatisfying because they leave us without a sense of the significance of what is being understood. We run the danger of crowding out meaning in search of the truth. Conversely, a gripping tale may reveal the significance of what goes on in the world at the expense of an accurate representation of the way things work. We can find deep meaning in patterns of fictional events. And it can be tempting to interweave or replace rudimentary explanations with meaningful stories in order to plug gaps in our understanding. But doing so frustrates the explanatory pursuit of truth. It is difficult to make adequate room for both sides of our nature, and all too easy for one aspect of human curiosity to crowd out the other. This danger is most pronounced in the case of topics that grip us deeply. Near-death experiences, we contend, is just such a topic.

So far, this book has focused on explanation, in particular, scientific explanation. We have been concerned with this pursuit, which aims at the truth, and how it can help us to make sense of near-death experiences. And we have contended that the prospects are better for coming to understand these experiences within the physicalist framework of scientific explanation than they are for making sense of them in supernatural terms. Now we want to consider the limits of scientific explanation and how some of the significance of near-death experiences can be made sense of in narrative terms. Our aim in doing so is the same as it has been. We want to make the case that coming to grips with near-death experiences does not require abandoning physicalism. Our main contention will be that the stories we tell about these experiences, and even our impulse to tell such stories, are wholly compatible with physicalism.

The basic idea behind our impulse to make sense of the world by means of stories is that placing events into a narrative form helps to render them meaningful for us.[5] Certain patterns of events are intelligible to us in a special way that allows them to have a distinctive kind of meaning. They resonate. Our understanding of

---

5. Here we are drawing on Velleman (2003) and (2009). See also Fischer (2009, Chs. 9 and 10) for further discussion of the role of narrative. Velleman's concern with narrative centers on issues about practical reasoning, and Fischer is concerned with issues about value, such as the value of lives and the value of acting freely. Our concern here is slightly different. We focus on the importance of narrative form for finding meaning in experience.

these sequences of events is not the same as the understanding we achieve through scientific explanation. Narrative understanding does not so much satisfy our yearning for an intellectual grasp of how things fit together. It satisfies our emotional sensitivity. Stories allow the world to become meaningful for us as creatures with feelings. They afford us a sentimental and often very deep grasp of what's around us. And they do this by placing sequences of events into patterns that grip our affective sensibilities. We feel the rise and fall of the narrative as it unfolds. Think of the first time you read *Romeo and Juliet*. Recall your anxious excitement as the young lovers see each other for the first time, feverish expectation at their courtship, crushing heartache at their deaths.

Crucially for our earlier claims about the similarity between the underpinnings of the significance of a near-death experience and an LSD trip, the meaning provided by a narrative grasp of things does not depend on an assumption that the events being placed in a narrative frame are real. Fiction can be meaningful, even fiction understood as such. We connect with made-up characters and their lives. And we do so knowing full well that they are figments of someone's imagination—projections in our own imaginations. The key to grasping the meaning provided by a narrative is the *form* that is taken by the events represented in that story. The connection between those events and an independent reality are beside the narrative point.

The moment Juliet stabs herself with the dagger is as inevitable as the moment Romeo drinks the poison. It just had to be that way. And yet we feel a certain kind of release, we achieve a sense of closure when the action of the play sweeps across the inevitable shores of the star-crossed lovers' deaths. We know where we are going, we cringe at the thought of the cruel fate that awaits our gaze, and yet we cannot help but feel relieved when we get there. The peace between their families is an afterthought. It seems inevitable as well, but no more than an elixir to dull the pain in our hearts. And yet that pain is the point. It ties everything together. The story of Romeo and Juliet, the story of stymied love throughout the ages, is meaningful precisely because it ends in sorrow and pain. It makes sense because we *feel* the predictable, foreseeable twists and turns of the plot as it unfolds. We ride the highs of the balcony and suffer the lows of the crypt because that is what it takes to achieve closure. Otherwise, none of this crazy, mixed-up dramatic world makes sense.

Everyone knows a Romeo and a Juliet. But they are not actually Romeo and Juliet. They do not really climb up on balconies and spurn Paris. The world gets in the way of their love, all right, but it does not end in deadly confusion for them. They mourn and move on, as people do. We console them and help them along their way. And yet we grasp the significance of what they are going through because we are familiar with the tale. It begins with a glimpse that gets the blood pumping and ends with heartache. The pattern is familiar. Sometimes it hits us that life is like a story.

Stories render events meaningful, thereby allowing us to comprehend them in a distinctive way. There is more than one form a sensible narrative can take. It can be one of high hopes dashed, of hard work rewarded, of the comeuppance that is one's due. What unites the various forms stories can take is that they allow us to fit sequences of events into familiar *emotional patterns*.[6] The important thing with a story is not how it describes reality—that is certainly not what we want in a good tale—but how it makes us feel. Stories afford a glimpse into the meaning of events by allowing us to wrap our hearts around them.

---

6. Here we are, once again, drawing on Velleman's insightful work.

This view about narrative and meaning allows us to understand how the LSD trip described earlier in this chapter may be meaningful in precisely the same way as a near-death experience. The trip contained the same events—a life review, an out-of-body experience, traveling through a tunnel of light, pleasant feelings, deceased people—fit into the same pattern as they might be in the context of a typical near-death experience. The LSD trip matches a typical near-death experience in both *form* and *content*. Given the view that the form of an experience is what allows it to be meaningful, this trip may be just as meaningful as a near-death experience. Of course, the elements of the LSD trip are *illusions*. And the subject of that experience takes them to be such. This marks a difference between his experience and the typical near-death experience. But this difference should not be taken to preclude the possibility that his hallucinatory experience may be meaningful for him in the same way as a near-death experience may be meaningful for someone else. Stories are meaningful because of the patterns in which they present events. They resonate with us because of how they make us feel. Fictional events can make us laugh and cry. There is no reason to think that a hallucinatory experience cannot be deeply meaningful as well.

One can keep in mind that one is tripping on LSD, and yet this experience can be deeply meaningful because of the emotional attunement one has in virtue of the pattern of events included in the experience. This suggests that one can react with awe and wonder in light of a physicalist understanding of what is going on in the case of near-death experiences. It would seem that one can keep in mind that one's near-death experience is best explained in terms of one's brain chemistry and psychology without losing one's emotional grip on the whole thing. One can have a deeply meaningful experience understood in physical terms because the explanation of the experience is a separate matter from the form of its contents. Thus, our contention that physicalist-friendly explanations of near-death experiences are better than supernatural explanations does not necessarily threaten the awe-inspiring power of these experiences.

This is not to say that focusing on the physical factors that account for why one had a given experience will never diminish its meaning or undermine feelings of awe and wonder. The point we are making is that there is good reason to think that these experiences may continue to be awe-inspiring even if they have physical explanations. Human beings, by our natures, seek explanation. But also, by our natures, we tell—and live—stories. The explanations need not crowd out the stories; indeed, they are both aspects of a more general project of comprehending the world and our place in it.

Moreover, we think that people are interested in *true* explanations. They want to understand why things really do occur. And from the perspective of this interest, it is neither here nor there whether a given explanation might affect one's reaction to the experience being explained. If true explanations undermine the meaningfulness of what is being made sense of, this does not show them to be any less true. Sometimes our interests come into conflict.[7] We want to make sense of the world, but we want to find meaning in it as well. When these two pursuits are incompatible, we face the difficult question of which side of our nature to give in to. However, to give in to our interest in meaning is not to change the terms in which we engage in explanation. It is to abandon that pursuit. To choose to preserve the significance of near-death experiences

7. Velleman (2009, Ch. 7) makes some very interesting remarks about this conflict in the context of practical reasoning.

over a true understanding of why they occur is to take sides with one aspect of human curiosity over the other. Our interest in true explanations remains intact, but swept to the side. In other words, it may still be the case that the best explanations of near-death experiences are in wholly physical terms, even if we decide that we are less interested in making sense of these experiences than we are in finding awe and wonder in them.

Thus far, we have argued for the claim that awe at near-death experiences is compatible with explaining them in physical terms. But what about hope? Many people find great consolation in the interpretation of near-death experiences as pointing to the existence of an afterlife. One might think that it is curmudgeonly (or worse) to advocate a view that threatens the hope afforded to so many on the basis of a supernatural understanding of near-death experiences.

The first thing to say in reply is that no part of our view in any way denigrates religious belief or a belief in the afterlife. We fully recognize and respect religious beliefs, and we are deeply cognizant of the hope that religion, and the doctrine of the afterlife, in particular, offers to many. We want to emphasize that nothing we have written is meant to dismiss or lead to a rejection of religious views or views about an afterlife. To be clear, our aim has been to call into question a *particular route* to religious beliefs and beliefs about the afterlife, namely, one that appeals to near-death experiences as evidence for, or even proof of, the reality of the afterlife. We have tried to show that this is not a compelling line of thought. But our position is fully compatible with accepting that there is, in fact, an afterlife. One could agree with everything we have said and yet still hold that heaven is for real. One might believe in the afterlife for reasons that have nothing to do with near-death experiences. The vast majority of religious people do in fact accept religion and hold views about the afterlife without basing these convictions on near-death experiences. Thus, our view is compatible with fully embracing the hope that religion can offer in the face of death.

People yearn for *genuine* hope—that is, hope based in reality and on good evidence—and *not* false hope, tempting as it may be. Belief on the basis of bad evidence and spurious arguments is weak and unstable. One way to strengthen your convictions is to reflect on your grounds for holding them in light of the most penetrating challenges. By grappling with the best arguments for opposing views, your own convictions can be rendered deeper and more secure. On recognizing that they are unsupported, you might purge evidently false beliefs. But the process of scrutinizing the grounds for your convictions need not result in changing your beliefs; rather, you may gain a deeper understanding of what you already believed to be true and why. By coming to grips with the best arguments against one's beliefs, you arm yourself with good ammunition for defending them to others and to yourself in times of doubt.[8]

Our aim has not been to change religious people's minds about fundamental reality. We are *not* aiming to make an atheist of you. Rather, we hope to have encouraged honest, thoughtful, and careful inquiry into a host of issues relevant to religious convictions central to many people's lives. We will have achieved our aim if we have gotten you to think critically about the arguments we oppose. It is all well and good, as far as we are concerned, if the end result is that you have a renewed commitment to the reality of the afterlife for some other reason than that near-death experiences seem to

8. See Mill (1859/1962: 180–181) for discussion of some of these points.

support it. Of course, we would be just as happy if our arguments have provided you with new and improved grounds for thinking that supernaturalism is false.

Once again, we are convinced that what people really want is *genuine* hope—hope based in *true* explanations, not wishful thinking. Unfortunately, even false hope has its allure. Hope is powerful and wonderful, and we can be tempted by it into accepting explanations we would not otherwise take seriously. Consider this passage from the chapter of *Map of Heaven* entitled, "The Gift of Hope," in which Alexander relates an account from Roger Ebert's wife of her husband's last days before succumbing to cancer:

> The one thing people might be surprised about—Roger said that he didn't know if he could believe in God. He had his doubts. But toward the end, something really interesting happened. That week before Roger passed away, I would see him and he would talk about having visited this other place. I thought he was hallucinating. I thought they were giving him too much medication. But the day before he passed away, he wrote me a note: "This is all an elaborate hoax." I asked him, "What's a hoax?" And he was talking about this world, this place. He said it was all an illusion. I thought he was just confused. But he was not confused. He wasn't visiting heaven, not the way we think of heaven. He described it as a vastness that you can't even imagine. It was a place where past, present, and future were happening all at once.[9]

9. Alexander (2014: 125–126).

Roger's note gave his wife, Chaz, hope, and this is a lovely thing. But now consider Alexander's analysis of the incident:

> It's fascinating, and always deeply moving to me, how people on the verge of leaving the world can—often after long and terrible suffering—suddenly catch a glimpse of where they are going, and of where they have actually been the whole time they were here. Ebert, a man who had made his living by words, wrote his wife a few words giving her what I am sure he felt was the most valuable gift he could possibly leave her: the truth about this world. Ebert is right. This world *is* an illusion. It's not real. And yet of course at the same time it *is* real, and wonderful, and deserving of our deepest love and attention. We just must not forget that it is not all there is.[10]

Alexander's comments here bring to mind the lessons we discussed in the previous chapter, regarding the human propensity toward confirmation bias. It involves gravitating toward confirming evidence and giving it a stronger and more dramatic interpretation than is strictly justified by the data itself. This is surely going on in Alexander's analysis of the episode involving the Eberts.

Note, first, that Mrs. Ebert reports that Roger was taking a significant amount of medication—so much that she worried that it was too much, just as she also worried about hallucinations, a common side effect of pain medication, especially in older patients. This is an immediate tip-off that something more mundane than contact with a more fundamental reality might be going on here. Moreover, we should pause to consider what exactly Roger might

---

10. Alexander (2014: 126).

have meant when he called our world "a hoax" and "an illusion." One might reasonably suppose that an individual at the end of his or her life might say this sort of thing to convey various different ideas, including that our ordinary preoccupations (with status, money, and so forth) are all fundamentally misguided. This is a plausible interpretation, but, of course, it has nothing to do with supernaturalism. Alexander interprets Roger to be saying that we will exist—and always have existed—in a realm that is somehow more "real" than our ordinary world—a realm which is left without description except that "past, present, and future were all happening at once." Alexander finds it fascinating and deeply moving that people at the end of their lives often "catch a glimpse of where they are going, and of where they have actually been the whole time they were here."

But this is, frankly, to give a spin on the episode that goes way beyond what is justified by the events related by Chaz Ebert. All we know is that a man who is being given a significant amount of medication (presumably, pain medication) after a long and terrible illness has expressed the thought that our lives are in some way illusory and has had experiences that apparently had content that suggested "vastness" and some sort of mixture of past, present, and future. Alexander's take is simply an over-interpretation of the data. Of course, it is not a surprising conclusion for Alexander to reach, as it would support his deeply held beliefs about the nature of the afterlife. This seems to be a stark instance of confirmation bias. The dependence of Alexander's interpretation on his prior beliefs becomes all the more obvious when you seriously consider what his interpretation involves. Is it really so easy to make sense of the past, present, and future "all happening at once?" How could we understand our lives as being part of a realm in which everything happens at once? This is perplexing, to say the least. But there are other available interpretations of what happened in this case. Perhaps Roger Ebert had a kind of "mystical" experience in which time did not appear to pass. We can make sense of this experience without being forced to try to comprehend a reality that involves past, present, and future rolled into one. And we are not forced to conclude that Roger had a "glimpse" of another realm or reality. Under the influence of medication and as he neared death, Roger Ebert could have taken a different *perspective* on reality and may have had profound and enlightening experiences; but this does not imply that he left our world and entered (even briefly) another one.

In evaluating claims about the afterlife, it is sensible to keep in mind our psychological propensities, some of which compel us to leap to unsupported conclusions. It is also prudent to take advantage of the wisdom of those who have been considering these issues for many years. Let us then close with some insightful advice from the Dalai Lama. The context of these words is a scene in which he shares the stage with Eben Alexander. We have discussed Alexander's near-death experience throughout this book, and it has given hope to many, including, it would seem, to Alexander himself. But as the Dalai Lama makes clear, in our quest for hope, we should be careful to inquire about the source. When faced with an apparently incredible story, one that we very much want to be true because it is so pregnant with hope, we should take care to properly vet the messenger. This is precisely the advice the Dalai Lama gives to the audience there to hear him and Alexander:

Both are here to speak at the graduation ceremony of Maitripa College, a Buddhist college in Portland, Oregon. Alexander is slated to speak first, and when he begins, the Dalai Lama cocks his head in a quizzical way and peers at him through his thick glasses.

Alexander tells his story like he's told it so many times before, in his soft, southern, confident burr. He tells the audience about the wondrous realm he visited, about the all-powerful and all-loving God he encountered there, and about some of the lessons he's brought back to earth. He says that among those lessons is the fact that reincarnation is real, and that knowing death is only ever temporary has helped him understand how a loving God can permit so many "tragedies and hardships and hurdles in the physical realm." As he did a few months ago, when Gretchen Carlson asked him whether the dead schoolchildren from Newtown remembered their slaughter, he offers comfort and hope. "I came to see all of those hardships as gifts," he says, "as beautiful opportunities for growth."

The Dalai Lama is not a native English speaker, and when it's his turn to speak, he does so much less smoothly than Alexander, sometimes stopping and snapping his fingers when a word escapes him, or turning to his interpreter for help when he's really stuck. He is not using notes, and the impression he gives is that of a man speaking off the cuff. He opens with a brief discourse about the parallels between the Buddhist and Shinto conceptions of the afterlife, and then, after glancing over at Alexander, changes the subject. He explains that Buddhists categorize phenomena in three ways. The first category are "evident phenomena," which can be observed and measured empirically and directly. The second category are "hidden phenomena," such as gravity, phenomena that can't be seen or touched but can be inferred to exist on the basis of the first category of phenomena. The third category, he says, are "extremely hidden phenomena," which cannot be measured at all, directly or indirectly. The only access we can ever have to that third category of phenomena is through our own first-person experience, or through the first-person testimony of others.

"Now, for example," the Dalai Lama says, "his sort of experience."

He points at Alexander.

"For him, it's something reality. Real. But those people who never sort of experienced that, still, his mind is a little bit sort of…" He taps his fingers against the side of his head. "Different!" he says, and laughs a belly laugh, his robes shaking. The audience laughs with him. Alexander smiles a tight smile.

"For that also, we must investigate," the Dalai Lama says. "Through investigation we must get sure that person is truly reliable." He wags a finger in Alexander's direction. When a man makes extraordinary claims, a "thorough investigation" is required, to ensure "that person reliable, never telling lie," and has "no reason to lie."

Then he changes the subject, starts talking about a massive project to translate ancient Tibetan texts.[11]

Our aim is not to call Alexander's character into question and thereby discredit his account of his near-death experience.[12] As we have stressed, we are willing to take at face value the reports of those who claim to have had near-death experiences. But we agree with the Dalai Lama that, especially in a context where extraordinary claims are being made, the responsible thing to do is to inquire critically.

11. Dittrich (2014).
12. There do, however, appear to be legitimate questions about Alexander's credibility due to facts about his past, as uncovered in the reporting of Dittrich (2014).

The purpose of this book has not been to discredit those who share their near-death experiences with the rest of us. Rather, it has been to critically examine the purported *implications* of these profound experiences. Whereas others have precipitously embraced supernatural interpretations of near-death experiences, we have argued that there is good reason to try to fit our understanding of them into the worldview supported by the physical sciences. And we have made the case that this approach can be sensitive to the transformational nature and meaningful character of these experiences. In the end, we think we can have our cake and eat it too. We can gain an understanding of near-death experiences that fits with the best theories provided by the physical sciences while at the same time appreciating the profound significance of these experiences for those who have had them. In making sense of the world, we seek the *best* explanation. We aim for truth. But we also seek to be at home in the world, and so we tell stories. Here we aim for beauty—a beauty we might well find in nature.

REFERENCES

ABC News. 2005. Parents think boy is reincarnated pilot. *ABC News* archive. http://abcnews.go.com/Primetime/Technology/story?id=894217. Accessed 6 November 2013.

Alexander, Eben B. 2012a. *Proof of heaven: A neurosurgeon's journey into the afterlife*. New York: Simon & Schuster.

Alexander, Eben B. 2012b. Heaven is real: A doctor's experience with the afterlife. *Newsweek* archive. http://www.newsweek.com/proof-heaven-doctors-experience-afterlife-65327. Accessed 8 October 2012.

Alexander, Eben B. 2014. *The map of heaven: How science, religion, and ordinary people are proving the afterlife*. New York: Simon & Schuster.

American Foundation for the Blind. 2014. Cortical visual impairment, traumatic brain injury, and neurological vision loss. http://www.afb.org/info/living-with-vision-loss/eye-conditions/cortical-visual-impairment-traumatic-brain-injury-and-neurological-vision-loss/123. Accessed 6 May 2014.

Bacon, Francis. 1620/2004. *The new organon*. In *The instaurato magna, part II: Novum organum and associated texts*, trans. Graham Reese and Maria Wakely. Oxford: Clarendon Press.

Ball, Philip. 2014. Beauty ≠ truth. *Aeon*. http://aeon.co/magazine/philosophy/beauty-is-truth-theres-a-false-equation/. Accessed 25 November 2014.

Becker, Ernest. 1973. *The denial of death*. New York: Simon and Schuster.

Blackmore, Susan J. 1993. *Dying to live*. New York: Prometheus Books.

Blaikie, Andrew. 2014. Visually impairing conditions (written for Scottish Sensory Centre). http://www.ssc.education.ed.ac.uk/resources/vi&multi/eyeconds/cereVI.html. Accessed 6 May 2014.

Borjigin, Jimo, UnCheol Lee, Tiecheng Liu, Dinesh Pal, Sean Huff, Daniel Klarr, Jennifer Sloboda, Jason Hernandez, Michael M. Wang, and George A. Mashour. 2013. Surge of neurophysiological coherence and connectivity in the dying brain. *Proceedings of the National Academy of Sciences of the United States of America* 110(35): 14432–14437.

Bortolotti, Lisa. 2010. Agency, life extension, and the meaning of life. *Monist* 93(1): 38–56.

Bowman, Carol. 2010. Children's past life memories and healing. *Subtle Energies and Energy Medicine* 21(1): 39–58.

Brennan, Zoe. 2009. Reincarnated! Our son is a World War II pilot come back to life. *Daily Mail Online* archive. http://www.dailymail.co.uk/femail/article-1209795/Reincarnated-Our-son-World-War-II-pilot-come-life.html. Accessed 6 November 2013.

Burke, Brian L., Andy Martens, and Erik H. Faucher. 2010. Two decades of terror management theory: A meta-analysis of mortality salience research. *Personality and Social Psychology Review* 14(2): 155–195.

Burpo, Todd, with Lynn Vincent. 2010. *Heaven is for real: A little boy's astounding story of his trip to heaven and back*. New York: Thomas Nelson.

Ceci, S. J., and M. Bruck. 1993. Suggestibility of the child witness: A historical review and synthesis. *Psychological Bulletin* 113(3): 403–439.

Ceci, S. J., and M. C. Huffman. 1997. How suggestible are preschool children? Cognitive and social factors. *Journal of the American Academy of Child & Adolescent Psychiatry* 36(7): 948–958.

Chabris, Christopher, and Daniel Simons. 2014. Why our memory fails us. *New York Times* archive. http://www.nytimes.com/2014/12/02/opinion/why-our-memory-fails-us.html. Accessed 2 December 2014.

Cook, Emily Williams, Bruce Greyson, and Ian Stevenson. 1998. Do any near-death experiences provide evidence for the survival of human personality after death? Relevant features and illustrative case reports. *Journal of Scientific Exploration* 12(3): 377–406.

Daston, Lorraine. 2014. Wonder and the ends of inquiry. *Point*. http://thepointmag.com/2014/examined-life/wonder-ends-inquiry. Accessed 15 December 2014.

Dean, Michelle. 2015. The boy who didn't come back from heaven: Inside a bestseller's "deception." *Guardian* archive. http://www.theguardian.com/books/2015/jan/21/boy-who-came-back-from-heaven-alex-malarkey. Accessed 22 January 2015.

Descartes, René. 1649/1989. *The passions of the soul*, trans. Stephen Voss. Indianapolis, IN: Hackett.

Dittrich, Luke. 2014. The prophet. *Esquire* archive. http://www.esquire.com/features/the-prophet. Accessed 6 May 2014.

European Union: The Human Brain Project. https:// www.humanbrainproject. eu/.

Ffytche, Dominic H. 2007. Visual hallucinatory syndromes: Past, present, and future. *Dialogues in Clinical Neuroscience* 9(2): 173–189.

Ffytche, Dominic H., R. J. Howard, M. J. Brammer, A. David, P. Woodruff, and S. Williams. 1998. The anatomy of conscious vision: An fMRI study of visual hallucinations. *Nature Neuroscience* 1(8): 738–742.

Fischer, John M. 2009. *Our stories: Essays on life, death, and free will.* New York: Oxford University Press.

Gabbard, Glen O., and Stuart W. Twenlow. 1984. *With the eyes of the mind: An empirical analysis of out-of-body studies.* New York: Praeger.

Greenberg, Jeff, Tom Pyszczynski, and Sheldon Solomon. 1986. The causes and consequences of a need for self-esteem: A terror management theory. In *Public self and private self,* ed. R. F. Baumeister, 189–212. New York: Springer-Verlag.

Greyson, Bruce, Emily Williams Kelly, and Edward F. Kelly. 2009. Explanatory models for near-death experiences. In *The handbook of near-death experiences,* ed. Janice Miner Holden, Bruce Greyson, and Debbie James, 213–234. Santa Barbara, CA: Praeger.

Holden, Janice Miner. 2009. Veridical perception in near-death experience. In *The handbook of near-death experiences,* ed. Janice Miner Holden, Bruce Greyson, and Debbie James, 185–212. Santa Barbara, CA: Praeger.

Katz, J., Grosman-Saadon, N., and Arzy, S. (ms.) The life-review experience: Qualitative and quantitative analysis.

Kelly, E. W., Bruce Greyson, and Ian Stevenson. 1999–2000. Can experiences near death furnish evidence of life after death? *Omega* 40(4): 513–519.

Leininger, Bruce, and Andrea Leininger, with Ken Gross. 2009. *Soul survivor: The reincarnation of a World War II fighter pilot.* New York: Grand Central.

Loftus, Elizabeth. 1997. Creating false memories. *Scientific American* 277(3): 70–75.

Long, Jeffrey, with Paul Perry. 2010. *Evidence of the afterlife: The science of near death experiences.* New York: HarperCollins.

Mill, John Stuart. 1859/1962. On liberty. In *Utilitarianism and other writings,* ed. Mary Warnock, 88–180. New York: Meridian.

Milligan, Wes. 2004. The past life memories of James Leininger. *Acadiana Profile Magazine.* http://www.iisis.net/index.php?page=james-huston-james-leininger-reincarnaton-wes-milligan-acadiana-profile&hl=en_US. Accessed 6 November 2013.

Mitchell-Yellin, Benjamin, and John Martin Fischer. 2014. The near-death experience argument against physicalism: A critique. *Journal of Consciousness Studies* 21(7/8): 158–183.

Moody, Raymond. 1975. *Life after life.* New York: Mockingbird Books.

Morse, Melvin, with Paul Perry. 1991. *Closer to the light.* New York: Ivy Books.

National Institutes of Health. 2008. Leading causes of blindness. *NIH Medline Plus* 3(3): 14–15.

National Institutes of Health. 2013. http://www.nih.gov/science/brain/.

Nelson, Kevin. 2014. Near-death experience: Arising from the borderlands of consciousness in crisis. *Annals of the New York Academy of Sciences* 1330: 111–119.

Pierce, Charles Saunders. 1877/1997. The fixation of belief. In *Pragmatism: A reader*, ed. Louis Menand, 7–25. New York: Vintage Books.

Pollan, Michael. 2015. The trip treatment. *New Yorker*, 9 February: 36–47.

Posner, Richard A. 2003. *Public intellectuals: A study of decline*. Cambridge, Mass.: Harvard University Press.

Robinson, Howard. 2012. Dualism. In *The Stanford Encyclopedia of Philosophy*, Winter 2012 edition. http://plato.stanford.edu/archives/win2012/entries/dualism. Accessed 6 May 2014.

Sacks, Oliver. 2012a. *Hallucinations*. New York: Alfred A. Knopf.

Sacks, Oliver. 2012b. Seeing God in the third millennium. *Atlantic* archive. http://www.theatlantic.com/health/archive/2012/12/seeing-god-in-the-third-millennium/266134/. Accessed 6 May 2014.

Sacks, Oliver. 2015. *My Life. New York Times* archive. http://www.nytimes.com/2015/02/19/opinion/oliver-sacks-on-learning-he-has-terminal-cancer.html?_r=0. Accessed 19 February 2015.

Schacter, Daniel L. 1987. Implicit memory: History and current status. *Journal of Experimental Psychology: Learning, Memory, and Cognition* 13(3): 501–518.

Shroder, Tom. 2014. *LSD, ecstasy, and the power to heal*. New York: Blue Rider Press.

Smit, Rudolf H., and Titus Rivas. 2010. Rejoinder to response to "Corroboration of the dentures anecdote involving veridical perception in a near-death experience." *Journal of Near-Death Studies* 28(4): 193–205.

Solomon, Sheldon, Jeff Greenberg, and Tom Pyszczynski. 2015. *The worm at the core: On the role of death in life*. New York: Random House.

Telegraph. 2009. Is James Leininger reincarnation of Second World War fighter pilot? *Telegraph* archive. http://www.telegraph.co.uk/news/newstopics/howaboutthat/6061466/Is-James-Leininger-reincarnation-of-Second-World-War-fighter-pilot.html. Accessed 6 November 2013.

van Fraassen, Bas. 1980. *The scientific image*. Oxford: Clarendon Press.

van Lommel, Pim. 2010. *Consciousness beyond life: The science of the near-death experience*. New York: HarperCollins.

van Lommel, Pim. 2013. Non-local consciousness: A concept based on scientific research on near-death experiences during cardiac arrest. *Journal of Consciousness Studies* 20(1–2): 7–48.

van Lommel, Pim, Ruud van Wees, Vincent Meyers, and Ingrid Elfferich. 2001. Near-death experiences in survivors of cardiac arrest: A prospective study in the Netherlands. *Lancet* 358: 2039–2045.

Velleman, J. David. 2003. Narrative explanation. *Philosophical Review* 112(1): 1–25.

Velleman, J. David. 2009. *How we get along.* New York: Cambridge University Press.

Vyse, Stuart A. 1997. *Believing in magic: The psychology of superstition.* Oxford: Oxford University Press.

Warren, A., K. Hulse-Trotter, and E. C. Tubbs. 1991. Inducing resistance to suggestibility in children. *Law and Human Behavior* 15(3): 273–285.

Wettstein, Howard. 2012. *The significance of religious experience.* New York: Oxford University Press.

Williams, Kevin. 2014. People have near-death experiences while brain-dead. http://www.near-death.com/experiences/evidence01.html. Accessed 12 May 2014.

Woerlee, G. M. 2010. Response to "Corroboration of the dentures anecdote involving veridical perception in a near-death experience." *Journal of Near-Death Studies* 28(4): 181–191.

Woerlee, G. M. 2011. Could Pam Reynolds hear? A new investigation into the possibility of hearing during this famous near-death experience. *Journal of Near Death Studies* 30(1): 3–21.

World Health Organization. 2014. Causes of blindness and visual impairment. http://www.who.int/blindness/causes/en. Accessed 6 May 2014.

INDEX

Note: "near-death experience" is abbreviated as "NDE"; "*n*" denotes notes.